Structural Integrity

AF844832

Volume 15

The *Structural Integrity* book series is a high level academic and professional series publishing research on all areas of Structural Integrity. It promotes and expedites the dissemination of new research results and tutorial views in the structural integrity field.

The Series publishes research monographs, professional books, handbooks, edited volumes and textbooks with worldwide distribution to engineers, researchers, educators, professionals and libraries.

Topics of interested include but are not limited to:

- Structural integrity
- Structural durability
- Degradation and conservation of materials and structures
- Dynamic and seismic structural analysis
- Fatigue and fracture of materials and structures
- Risk analysis and safety of materials and structural mechanics
- Fracture Mechanics
- Damage mechanics
- Analytical and numerical simulation of materials and structures
- Computational mechanics
- Structural design methodology
- Experimental methods applied to structural integrity
- Multiaxial fatigue and complex loading effects of materials and structures
- Fatigue corrosion analysis
- Scale effects in the fatigue analysis of materials and structures
- Fatigue structural integrity
- Structural integrity in railway and highway systems
- Sustainable structural design
- Structural loads characterization
- Structural health monitoring
- Adhesives connections integrity
- Rock and soil structural integrity.

**** Indexing: The books of this series are submitted to Web of Science, Scopus, Google Scholar and Springerlink ****

This series is managed by team members of the ESIS/TC12 technical committee.

Springer and the Series Editors welcome book ideas from authors. Potential authors who wish to submit a book proposal should contact Dr. Mayra Castro, Senior Editor, Springer (Heidelberg), e-mail: mayra.castro@springer.com

More information about this series at http://www.springer.com/series/15775

Grzegorz Lesiuk · José A. F. O. Correia ·
Halyna V. Krechkovska · Grzegorz Pekalski ·
Abílio M. P. de Jesus · Oleksandra Student

Degradation Theory of Long Term Operated Materials and Structures

Springer

Grzegorz Lesiuk
Faculty of Mechanical Engineering
Wroclaw University of Science and Technology
Wrocław, Poland

Halyna V. Krechkovska
Karpenko Physico-Mechanical Institute
Lviv, Ukraine

Abílio M. P. de Jesus
Faculty of Engineering
Deptartment of Mechanical Engineering
University of Porto
Porto, Portugal

José A. F. O. Correia
Faculty of Engineering
University of Porto
Porto, Portugal

Grzegorz Pekalski
Wroclaw University of Science and Technology
Wrocław, Poland

Oleksandra Student
Karpenko Physico-Mechanical Institute
Lviv, Ukraine

ISSN 2522-560X ISSN 2522-5618 (electronic)
Structural Integrity
ISBN 978-3-030-43712-1 ISBN 978-3-030-43710-7 (eBook)
https://doi.org/10.1007/978-3-030-43710-7

This Springer imprint is published by the registered company Springer Nature Switzerland AG
The registered company address is: Gewerbestrasse 11, 6330 Cham, Switzerland

Preface—Degradation Theory of Long Term Operated Materials and Structures

The maintenance and reliability of historical objects require precise diagnostic methods as well as technical condition assessment, often exceeding the procedures adopted in the standard regulations for modern engineering structures. It is particularly relevant for historical steel bridge structures engineered in the late nineteenth and early twentieth centuries. The diversion of operating conditions, as well as the unique features of historical materials, require an individual approach to assessing their technical condition and suitability for further use. For a wide range of historical bridge structures, it should be noted that the 19th and 20th-century steel structures were made of puddle iron and low-carbon old mild steel. In metallic materials from the turn of the 19th and 20th century, the tendency to microstructure degradation processes is observed. Overall changes in material microstructure as a result of long-term operation under dynamic loads and corrosive threats resulting in a change (decrease) in strength properties. This phenomenon was called "material aspects of degradation theory".

In many publications, this term is associated with "ageing of the material", but in the Authors' opinion, these processes are not comparable. However, the term ageing does not have any negative connotations. As a heat treatment procedure, it leads to an increase in the strength properties of materials. Degradation changes in low carbon constructional steel microstructure consist of separation processes of carbides and iron nitrides inside the grains, formation of carbide envelopes at grain boundaries. The microstructural corrosion changes can support the mentioned effects and finally lead to the initiation and growth of micro-cracks in materials. Therefore, this is a different concept from the term "ageing".

In recent years, there has been an increasing number of publications on methods of diagnosis and methods of assessing the technical condition of age-advanced objects. Therefore, the authors of this publication undertook to evaluate the impact on the strength and explanation of the processes occurring in the microstructure of long-term operated nineteenth-century (and not only) materials. The knowledge about the intensity and direction of degradation processes development can improve the selection of the proper diagnostic methods (also described in this book). Obtained results may be also useful in the near future refereed to the objects erected in the '50s and '60s of the last century. They will probably require a comprehensive assessment of their technical conditions. In the context of the comprehension of processes that occur in the material, as well as prediction of directions and tendencies of their development and the impact on the material's behaviour under cyclic and dynamic loads. So, the need for experimental research and an explanation of the occurring processes theory is well justified. The obtained results and conclusions may constitute in the nearest future a starting point for considerations on the methodology of technical condition assessment of a slightly lower age steel structure. The book is addressed to maintenance engineers, academics and students as a source of information on the edge of mechanics, civil engineering and materials science. Besides, it is a unique collection of data and research results on historical constructional metallic materials and connections.

The authors would also like to express their gratitude to all the institutions that have financially (and not only) contributed to the book's preparation, i.e.:

- Wrocław Academic Hub (Wrocławskie Centrum Akademickie)—This book has been prepared as a part of the Support Programme MOZART of the Partnership between Higher Education and Science and Business Activity Sector financed by City of Wroclaw;
- Wrocław University of Science and Technology, Faculty of Mechanical Engineering, Department of Mechanics, Materials Science and Engineering;
- University of Porto, Faculty of Engineering—This research was funded by grant number POCI-01-0145-FEDER-030103 FiberBridge—Fatigue strengthening and assessment of railway metallic bridges using fiber-reinforced polymers by FEDER funds through COMPETE2020 (POCI) and by national funds (PIDDAC) through the Portuguese Science Foundation (FCT/MCTES). Additionally, this work was financially supported by: Base Funding—UIDB/04708/2020 and Programmatic Funding—UIDP/04708/2020 of the CONSTRUCT—Instituto de I&D em Estruturas e Construções—funded by national funds through the FCT/MCTES (PIDDAC).
- Karpenko Physico-Mechanical Institute of the National Academy of Sciences of Ukraine.

Wrocław, Poland	Grzegorz Lesiuk
Porto, Portugal	José A. F. O. Correia
Lviv, Ukraine	Halyna V. Krechkovska
Wrocław, Poland	Grzegorz Pekalski
Lviv, Ukraine	Oleksandra Student
Porto, Portugal	Abílio M. P. de Jesus
July 2020	

About This Book

The maintenance and reliability of historical materials and structures require precise diagnostic methods as well as technical condition assessment, often exceeding the procedures adopted in the standard regulations for modern engineering structures. This is particularly relevant for historical steel bridge structures engineered in the late nineteenth and early twentieth centuries. The increase of the service loading, as well as the unique features of historical materials, require an individual approach for assessing their technical condition and suitability for further use. Overall changes in material microstructure as a result of long-term operation under dynamic loads and corrosive threats resulting in a change (decrease) in strength properties have been experienced by old materials and structures that have been handled in this book within the background of the "material aspects of degradation theory". Additionally, this book addresses the fatigue behaviour of riveted connections from ancient Portuguese metallic bridges as well as the fatigue crack propagation behaviour of related materials. The book intends to be suitable for maintenance engineers, academics and students as a source of information on the edge of mechanics, civil engineering and materials science. Besides, it is a unique collection of data and research results on historical constructional materials and structural connections.

About This Book

Contents

About the Authors

Dr. Grzegorz Lesiuk Associate Professor in the Department of Mechanics, Materials Science and Engineering, Faculty of Mechanical Engineering at the Wrocław University of Science and Technology, Poland. The research field of Dr. Lesiuk is fatigue & fracture of engineering materials connected with the microstructural degradation processes and its potential influence on the mechanical properties of long-term operated steel structural members. He is also interested in new materials, composite materials and mathematical modelling of fatigue & fracture process regarding energy approach. Dr. Lesiuk at Wroclaw University of Science and Technology conducts regular lectures and classes: classical mechanics, dynamics, differential equations, fracture mechanics, lightweight structures, engineering failure analysis. Dr. Lesiuk is a member of Polish Group of Fracture Mechanics, Polish Society of Theoretical and Applied Mechanics, ESIS (European Structural Integrity Society), co-chair of the working group in TC12 ESIS, member of the advisory board of Springer Book Series-Structural Integrity, member of editorial board Fracture and Structural Integrity Journal (FIS), Science Progress (SAGE). He is also an author and co-author of more than 150 scientific publications in the most relevant national and international scientific journals devoted to engineering materials and mechanics, proceedings in international and national conferences, congresses and workshops, author of 1 book.

José A. F. O. Correia is Researcher at CONSTRUCT/FEUP of the University of Porto (Portugal). Since 2018, he is a Guest Teacher at the Engineering Structures Department of the Civil Engineering and Geosciences Faculty of the Delft University of Technology (Netherlands). He is an invited Professor at the Structural Mechanics section in the Civil Engineering Department of the University of Coimbra. He has obtained his B.Sc. (2007) and M.Sc. (2009) degrees in Civil Engineering from the University of Trás-os-Montes e Alto Douro. He was a specialist in steel and composite (steel and concrete) construction at the University of Coimbra in 2010. He was awarded his Ph.D. in Civil Engineering from the University of Porto in 2015. He is also a co-author of more than 100 papers in the most relevant scientific journals devoted to engineering materials and structures as well as 200 proceedings in international and national conferences, congresses, and workshops. He is a member of scientific and professional organizations, such as Ordem dos Engenheiros, Associação Portuguesa de Construção Metálica e Mista (CMM), and Associação o para a Conservação e Manutenção de Pontes (ASCP, Portuguese Group of IABMAS). He is Co-Chair of TC12 of European Structural Integrity Society (ESIS), the Editor of the Springer Book Series Structural Integrity, and Guest Editor of numerous international journals. He is a member of technical committees, CEN TC250 SC3 WG 1–9 Fatigue and CEN TC250 SC3 WG 1–10 Fracture, of the European Committee for Standardization (CEN). His current research interests are (a) behaviour to fatigue and fracture of materials and structures (steel and aluminium, riveted and bolted connections, pressure vessels, old steel bridges, wind turbine towers, offshore structures); (b) probabilistic fatigue modelling of metallic materials (including statistical evaluation, size effect, cumulative damage); (c) probabilistic design of glass structural elements; and (d) mechanical behaviour of materials and wooden structures (connections and characterization of ancient structures).

Halyna V. Krechkovska, Ph.D. is Senior Research in the Department Diagnostics of Materials Corrosion-Hydrogen Degradation of the Karpenko Physico-Mechanical Institute of National Academy of Sciences of Ukraine. The scientific activity is mainly focused on the examination technical state of old constructions, determination of the causes of destruction of structure elements in the oil and gas production, heat power and oil refining industries of Ukraine. The development of approaches for the quantitative assessment of fractographic and structural features of steel degradation and established their correlation with its standard mechanical properties is within her scientific interests. She has over 83 published scientific papers and 4 patents. Since 2015 is the Member of European Structural Integrity Society (ESIS).

Docent Dr. Grzegorz Pekalski Emeritus Assistant Professor at the Faculty of Mechanical Engineering, Wrocław University of Science and Technology, Poland. The research field of Dr. Pękalski is materials science and corrosion issues. Docent Grzegorz Pękalski was responsible for the execution of several national projects related to the degradation of materials in the power industry and mining industry—especially brown coal excavators. He is the author of an innovative concept for the assessment of the materials degradation state. He was responsible for the education of the young generation of scientists at Wrocław University of Science and Technology-Faculty of Mechanical Engineering. Many of his students have been awarded for scientific achievements. He also contributed to the implementation of materials engineering in education programmes within the Individual Study Programme at Wroclaw University of Science and Technology.For his academic achievements, he was honoured many times by the Polish authorities. He is also an author and co-author of more than 200 scientific publications in the most relevant national and international scientific journals devoted to engineering materials, proceedings in international and national conferences, congresses and workshops, author of 3 books.

Abílio M. P. de Jesus, (Ph.D., M.Sc.) is Associate Professor at the Department of Mechanical Engineering of the Faculty of Engineering from the University of Porto (FEUP) since 2014, after 18 years of teaching activity at the Department of Engineering of the University of Trás-os-Montes e Alto Douro (UTAD), Vila Real, Portugal. He is also a research member at the Institute of Science and Innovation in Mechanical and Industrial Engineering (INEGI) and is integrated research in the Associated Laboratory for Energy, Transports and Aeronautics (LAETA) and responsible for the Thematic Research Line on Advanced Manufacturing. He graduated in Mechanical Engineering from FEUP in 1996. He got his master's degree in Mechanical Engineering from FEUP in 1999. He got a Ph.D. degree in Mechanical Engineering at UTAD in 2004. He is co-author of 160 papers in national and international scientific journals and more than 300 papers presented and/or published into proceedings of both national and international conferences. A total of 206 documents in SCOPUS and an H-index of 24 show the consistency and visibility of the research work that has been developed in the fields of fatigue and fracture of materials and structures. The manufacturing processes, and particularly subtractive and additive processes, have been elected as a research priority in the last 5 years with a group in the field consolidated. Process simulation, material characterization, hybrid manufacturing (additive and subtractive) and their relation with fatigue behaviour are targeted besides other research topics related to the integrity of structures of different applications. The edition of the Structural Integrity Book series, the nomination for the ESIS TC12 and the current European Projects, FASTCOLD and IN2TRACK2, and national projects MAMTOOL, ADD.STRENGTH and FIBREBRIDGE highlights the research being performed.

Prof. Oleksandra Student is a leading researcher in the Department of Diagnostic of Hydrogen-Corrosion Degradation of Materials, Karpenko Physico Mechanical Institute of the National Academy of Sciences of Ukraine. The main directions of scientific activity: fractographic examination of damaged structural elements in the oil and gas production and transportation systems, heat power and oil refining industries; the assessment of the technical state of the structural steels after its long-time in-service degradation using a wide range of mechanical characteristics. Author a 3 monographs and 275 scientific papers, 20 patents.

Chapter 1
Introduction to the Degradation Theory of Low Carbon Steels

Abstract In this chapter, the problem of deterioration of the mechanical properties of metallic structural members is presented. This phenomenon is associated with microstructural changes in the long-term operated structural objects, especially heavy machinery and bridges. The basic mechanisms of degradation are presented and classified. Further, basic research and diagnostic methods are described. Next, the results of the own studies are presented and the degradation changes are illustrated by microscopic images. As it has been presented, in the case of long-term elements, the adopted test methods should aim at a detailed assessment of the degree of deterioration of mechanical properties as a result of the activity of degradation processes.

1.1 Introduction to the Degradation Theory

The concept of degradation is currently used in many contexts, including non-technical ones. In relation to the issues discussed in this chapter, they are defined in works [1, 2]. They were formulated with reference to long-lasting lignite mining machinery as follows: "*Degradation is a process of deterioration of strength and utility properties of elements, nodes, and thus of the whole object*". In an intuitive sense, degradation of technical objects is associated with the progressive loss of functional and functional properties. The paper [2] presents the main ideas and assumptions of the theory of degradation used to assess the usefulness of the object for further operations. The theory of degradation is constantly being developed and expanded, as evidenced by the works selected by the authors of this book: [3–10]. The concept of degradation is synonymous with the concept of ageing, caused by the operating time, expressed in years or the number of load cycles. In this sense, degradation is related to the influence of external factors, which are the working environment of the machines and structures:

$$S_{ag} = S_{ag}(t, \Omega) \tag{1.1}$$

G. Lesiuk et al., *Degradation Theory of Long Term Operated Materials and Structures*, Structural Integrity 15, https://doi.org/10.1007/978-3-030-43710-7_1

In Eq. (1.1) S_{ag} represents the natural ageing process determined with the influence of exogenous factors that create the working environment.

However, in real operating conditions it cannot be described by such a simple model (1.1). Degradation phenomena are much more complex and depend on the accumulation of various effects of wear and tear processes. According to the [2, 3] degradation can be characterized as follows:

$$D_{\deg} = D_{\deg}\left(t, \Omega : F_{fat}, F_{fra}, T_{tri}, D_{dis}, C_{cre}\right) \tag{1.2}$$

where:

F_{fat}—fatigue degradation processes;
F_{fra}—fracture degradation processes;
T_{tri}—tribological degradation processes;
D_{dis}—deformation degradation processes;
C_{cre}—creep degradation processes.

This is scientific instrumentation which allows predicting the effects of long-term or improper exploitation of a technical object [2, 3]. The theory of degradation is, therefore, a synthesis of specialized knowledge in various fields. Such facts, theories, hypotheses and empirical data are presented in a systemic way. The theory of degradation strives in its nature to answer the fundamental question of how slow changes in measurable (physical) parameters can lead to continuous (wear and tear) or rapid (failure) qualitative changes in the mechanical objects. Therefore, (according to works [2, 3]) the theory of degradation should be:

- focused on tangible, quantifiable factors and ignore or diminish those factors that are intangible and not quantifiable,
- do not to take into account such "subtle" values as, for example, environmental issues, or aesthetic,
- focused on experimental or analytical results and on the logical justification for this the means and methods of reasoning,
- strived for objectivity, using the precise language of mathematics, even in matters of strong moral or social colouring.

The theory of degradation, like any empirical theory, must first present in a comprehensible way the goals for which it is created. These will be primarily cognitive goals, concerning [2]:

- a scientific explanation of experimentally observed facts,
- acquisition and elaboration of the complex procedures of these facts in defined sub-sets knowledge,
- systematization of sub-collections the existing knowledge,
- creation of systems to control the truthfulness of knowledge. The System shall be responsible for the implementation of non-verifiable generalizations, detecting

and describing the relations between different areas of the real operating conditions. In combination with econometrics, it can be responsible for emphasizing the interdisciplinary nature of theory,
- formulating new problems and directions of research on the basis of these inconsistencies.

Relevant practical goals for engineering purposes are as follows:

- influencing the appropriate shaping of the creative environment (designers, constructors, manufacturers, operators) by:
  - improvement of existing—empirical and artificial—computing and operating systems,
  - optimizing, on this basis, the design, construction and operation systems.
- prediction of empirical facts not yet observed, which to some extent is also a cognitive function of the theory.

Achievement of these objectives is possible due to the structure of the technical base and a number of empirical procedures developed within the framework of the theory of degradation, a diagram of which is presented in the Fig. 1.1. The most complex problem in the general technical base and procedural empirical theory of degradation (Fig. 1.1) is, in the opinion of the authors of the paper [2], determination of loads on working machines. The lack of load history is particularly negative when estimating fatigue life. The assessment of the remaining working time of the structure using the methods of material fatigue science as well as cracking mechanics focuses mainly around two currents. The first is the deterministic approach—based on mathematical relations between the number of cycles and stress or strain. The second approach is to take a statistical view of the problem of assessing the remaining fatigue strength. It is based on hypotheses and statistical distributions. Determinism and indeterminism have always divided the world of researchers into two groups. Kazimierz Sobczyk writes about the need to reconcile these two independent approaches at work [11]: "*Recognizing and quantifying the phenomena of the world around us cannot be complete (and in many situations it is directly impossible) without taking into account the role of randomness. The randomness of events and phenomena has troubled people since the earliest times; its mysterious essence has been and still is the subject of interest in philosophy and methodology of science, and its regularities are discovered by mathematics and empirical sciences*". Determinism and randomness are an integral part of the theory of degradation. The probabilistic description is also richer than the individual description, especially if we consider the possibility of unstable situations. In such situations probability plays an inherent role [9]. Therefore, when recreating the history of the load, it seems justified, as suggested by the authors [3], to adopt a load model in the form of a statistical distribution. This distribution is based on a representative set of results from random, planned experimental studies. It should be noted that over the last 30 years since the publication [3], progress in the field of electronics and sensors enables permanent recording and

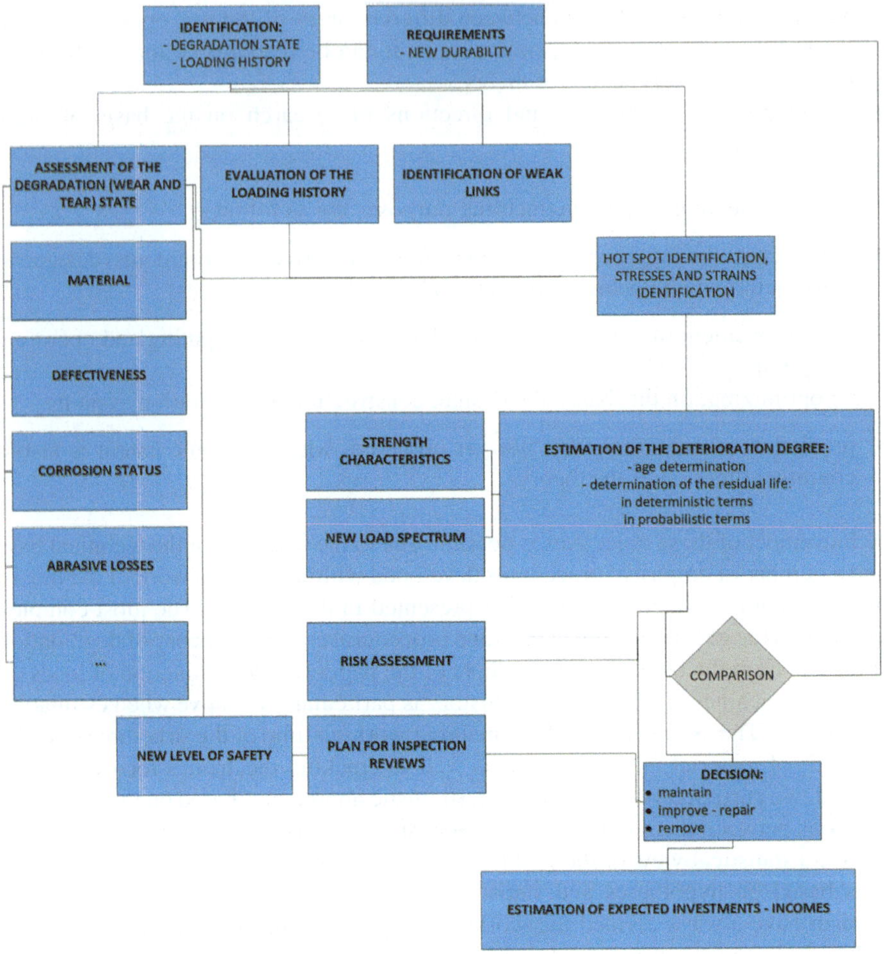

Fig. 1.1 The general technical structure of the degradation theory and its empirical procedures (based on [1–3])

processing of registered signals. However, in the case of age-old technical facilities, i.e. lignite excavators or bridges, there is still a need to approximate the history of the load with appropriate distributions due to the lack of a sufficiently large collection of information about the past of the facilities.

1.2 Material Aspects of the Degradation Theory

The studies on the theory of degradation [4–6, 12, 13] show the presence of microstructural degradation processes in structural steels. These changes are caused by the separation of metastable phases, consistent with the metallic matrix. The result is a deterioration in the plasticity of the steel and a tendency to brittle fracture. In addition, cyclic loading, as well as operation at elevated temperatures, intensify the development of degradation processes. This problem is particularly noticeable for steels subjected to long term, intensive stress and working conditions:

- in power engineering at elevated temperatures;
- in the hydrogen environment;
- with low carbon content.

Concern for the safety of such structures requires the determination of mechanical properties of the long-term operated steels, as well as determining the directions of development and the degree of severity of future degradation changes and their impact on mechanical properties. Therefore, the need to develop new diagnostic indicators enabling a reliable assessment of the condition of degradation is also clear. In the paper [13] it is possible to find the analysis and classification of degradation processes occurring in long-term operated steels. Typical symptoms of microstructural degradation are described as follows:

- progressive decomposition of perlite or bainite areas into ferrite and carbides;
- progressive secretion processes (nitrides and carbides);
- creation of local separation grids at the boundaries of ferrite grains;
- changes in the morphology of precipitations;
- changes in the dislocation structure of the material.

The phenomena mentioned above are conditioned by the chemical composition of the material and segregation. It should be noted that the degradation processes exit more often in unresolved steels than in deoxidized steels [13]. This is justified by the basic literature in the field of metallurgy. According to the study [14], steadily unquiet in comparison with deoxidized steels cause strong segregation of sulphur, phosphorus and carbon. The concentration differences between the surface and middle layers are up to 170% for carbon, up to 300% for phosphorus and up to 600% for sulfur. Permanently non-deoxidized steels have worse walkability and are more sensitive for structural ageing [14]. When considering one of the formula's components (Eq. 1.2) associated with the fatigue wear process—*{Ffat}*, it should be noted that, deformation wear reduces the fatigue strength, which in turn depends on the state of degradation of microstructures, and thus on the material wear associated with the deterioration of the strength properties of steel. As part of the theory of degradation and its material aspects, a multi-criteria system for solving material problems in the theory of machine degradation was developed [13]. The diagram of this system is presented in Fig. 1.1. The basis of the system is the modified (Fig. 1.2) structure of the technical base and the procedures of the empirical theory of degradation. The key problem is the select of nodes, i.e. the places where specimens can be extracted for testing. In the case of opencast mining machines, the choice of these sites is not obvious. In the paper [4] it was noted that the critical members are not always the members subjected to the maximal operating loads. Therefore, structural members and joints were selected

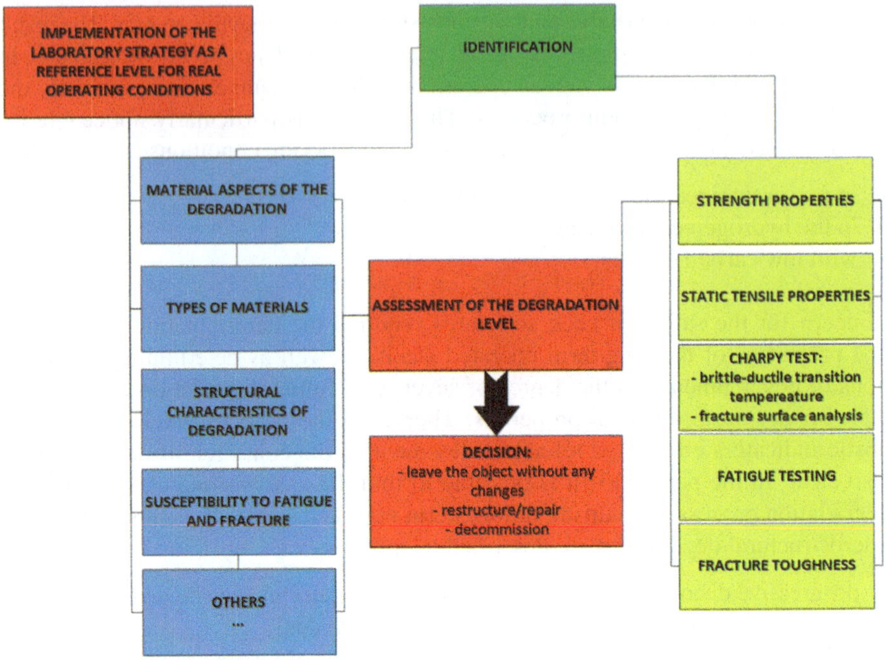

Fig. 1.2 Simplified multicriteria scheme supporting the maintenance engineering system

for testing because of operational data, taking into account their condition, they were considered critical and the most exposed to material wear and tear. Another problem is the state of the structure of the material at the time of its first application. For example, steels working in the power industry (after 10–20 years of operation period) do not cause any problems with identification and comparison with new materials due to small differences in the way of production of these steels. In the case of technical objects such as bridges or viaducts, the 100-year service life is already a significant discrepancy in initial and post-operated mechanical properties.

In the context of the development of steel manufacturing technology, it is not possible to accept as before the degradation state of the structure and properties of newly manufactured steels of similar (or the same) chemical composition. One solution would be to take the normalized (in terms of heat treatment procedure) state as a reference level in relation to state before degradation. This approach is justified by the use of an annealing procedure normalizing as a regenerative heat treatment method. Normalization allows considering the microstructure of the material as close as possible to the equilibrium state. The instrumentation of the theory of degradation and multi-criteria system of solving material problems assumes that degradation features can be detected by examination [13]:

- analysis of the microstructural changes identified by light and electron microscopy methods;

- the yield point and tensile strength and ductility features such as elongation at break and necking;
- impact strength and ductile-brittle transition temperature;
- fracture mechanics tests (fracture toughness, fatigue crack growth rate) in post-operated state and after normalizing heat treatment.

Specimens for the microstructure analysis and strength properties tests should be taken from cross-sections longitudinal and transverse orientation. Microscopic observations shall be made by the methods of light microscopy (LM) on non-etched and etched samples. Fractography study (after impact tests) is performed by means of an electron microscope scanning at magnifications of up to 3000x. As a result, the nature of the fractures is determined, the type, quantity and distribution of non-metallic inclusions and identification shall be carried out the components of the structure. Observations of thin films are carried out using the following methods transmission electron microscopy at magnifications of up to 200,000x in order to determine the type and degree of advancement of ageing processes, dislocation types and structures and the analysis of plastic deformations of the material caused by operating conditions. The application of electron diffraction methods enables the phase analysis of structure components appearing during ageing processes [12]. As an example and illustration of the mentioned facts, in the publications [4, 12, 13] the structure and strength properties of the degenerated structures were analyzed. The metallic materials were extracted from the structural components of lignite excavators SRs-1200. During the study the following symptoms of microstructural degradation were revealed of these steels:

- separation of tertiary cementite in the form of a thick envelope at the ferrite grain boundaries;
- separation of $Fe_{2\text{-}3}CN$ carbide-nitrides s in areas of ferrite grains causing the dislocation restraint, strengthening of ferrite and deterioration of its plastic properties;
- separation of $Fe_{16}N_2$ type nitrides inside ferrite grains.

Examples of such degraded structures illustrated by microscopic imaging methods—Figs. 1.3 and 1.4 [15].

Microstructural degradation changes in the form of separations are shown in Figs. 1.3 and 1.4. Cementite at the grain boundaries and numerous nitride and carbide secretions inside the grains are shown in Fig. 1.3b. Non-metallic inclusions are also noticeable (plastically deformed sulfides). Figure 1.3c shows the structure of S235JRG11 steel. (St3SX) after 55 years of operation. The structure of ferrite grains with perlite areas is noticeable with a large number of non-metallic inclusions and Fe_3C carbide precipitations at the grain boundaries. Figure 1.3 shows the structure of S235JRG1 (St3SX) steel after 43 years of operation after normalization. As it is expected, here is noticeable the structure of ferrite grains with a small amount of fine shallow perlite and tertiary cementite forming a continuous coarse envelope at the

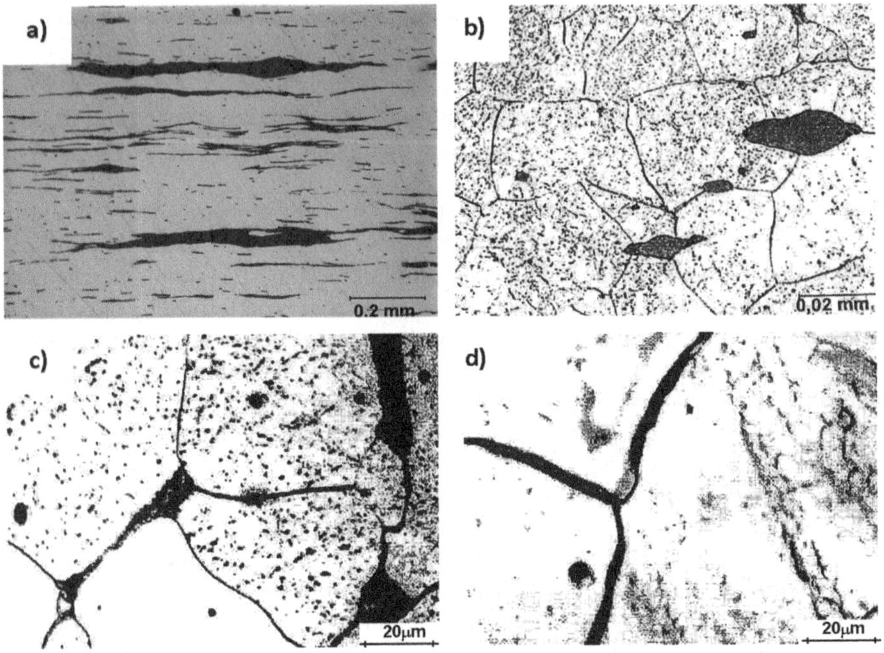

Fig. 1.3 Examples of microstructures of low-carbon steels: **a** microstructure of puddle iron non-etched, **b** degraded microstructure of ferritic steel, etched 3% HNO_3—ferrite grains (with plastically deformed sulphides) with numerous separations of nitrides and carbides, decorated with a continuous thick envelope of cementite at the grain boundaries, **c** steel S235JRG1 (St3SX) after 55 years of operation, 3% HNO_3—grain structure was etched ferrite with areas of perlite, brittle separation of Fe_3C carbide at ferrite grain boundaries, **d** S235JRG1 (St3SX) steel after 43 years of operation—normalized state, 3% HNO_3 etching—structure of ferrite grains with a small amount of fine-lamellar perlite and cementite forming a continuous envelope at the boundaries of the ferrite grains

grain boundaries. SEM and TEM images of the materials are shown in Fig. 1.4a–d. Figure 1.4a shows the microstructure of long-term operated steel in pos-operated condition, the following are visible ferrite grains with a minimum number of carbides and nitrides (A), as well as grains with a high percentage of nitrides (A) density of these separations (B), Fe_3C envelope at the borders (C), shallow-shaped perlite (D). Figure 1.4 shows the microstructure of old mild steel in post-operating condition with a few precipitations of cementite and nitrides (A). Numerous precipitations of nitrides were also depicted as well as carbides inside the ferrite grains (B), with C—thick envelope Fe_3C, and D—partially degenerated perlite. The TEM images are shown in Fig. 1.4c–d. Figure 1.4 you can see the microstructure of S235JRG1 (St3SX) steel after 43 years of operation—in the middle of parts, non-metallic inclusions with numerous dislocations are observed. Figure 1.4d shows the microstructure of S235JRG2 (St3SX) steel after 43 years of operation and in a normalized state.

As can be concluded from above, the research carried out with the use of light and scanning microscopy methods show the following significant diversification of

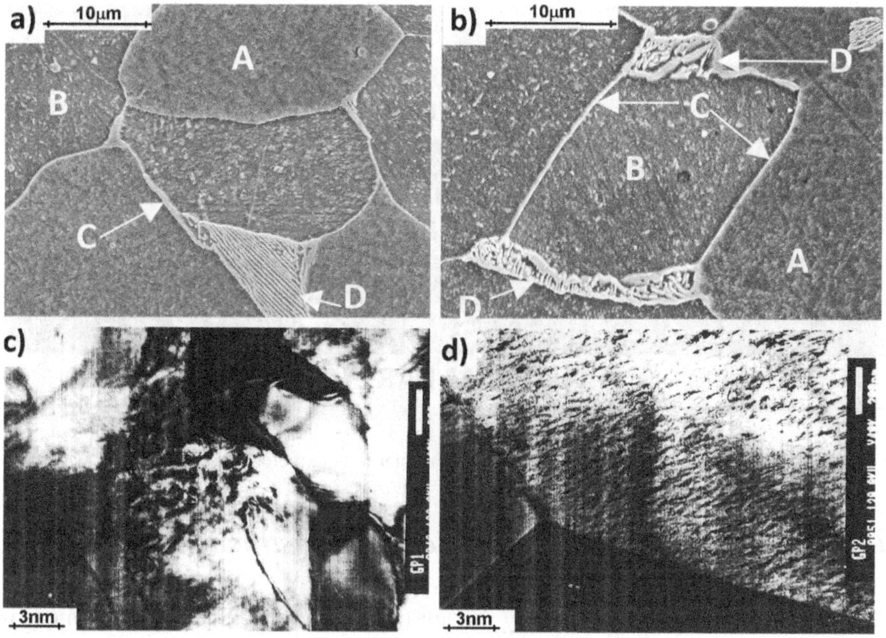

Fig. 1.4 Examples of microstructures of low-carbon steels—SEM (Scanning Electron Microscopy) and TEM (Transmission Electron Microscopy); **a**–**b** old mild steel after 100. years operating time; ferrite grains with a small amount of perlite, scanning microscopy, 3% HNO3 (description in the text), **c** steel S235JRG1 (St3SX) after 43 years of operation—in the central parts visible separation of non-metallic inclusions with dislocations, electron microscope transmission—thin foil, **d** S235JRG1 steel (St3SX) after 43 years of operation and annealing normalizing—ferrite grains with very fine, undissolved secretions of nitrides, electron transmission microscope—thin foil

structures of the examined materials. TEM observations allow identifying phases and structures. On the other hand, the use of SEM in fractographic research is extremely important, in case of the failure analysis of the specimens after Charpy impact tests—Fig. 1.5. The observations are usually carried out in three stages the characteristic zones of the specimens, i.e. close to mechanical notch (A), central part of the specimen (B), bottom part of the fractured area (zone C).

Examples of the fracture images after impact tests (SEM-registered) are shown in Fig. 1.6a–f. On the basis of the analysis, it can be concluded that in case of the tested steel significant differences fractured zones; A and C (obtained in the temperature range +20 °C and −40 °C) are noticeable. Low impact strength of the tested materials—e.g. specimen from Fig. 1.6d–f, broken at temperature −40 °C, showing the brittle nature of the fracture surface along the cross-section, which corresponds with low Charpy test result: KCV = 5.2 J/cm^2.

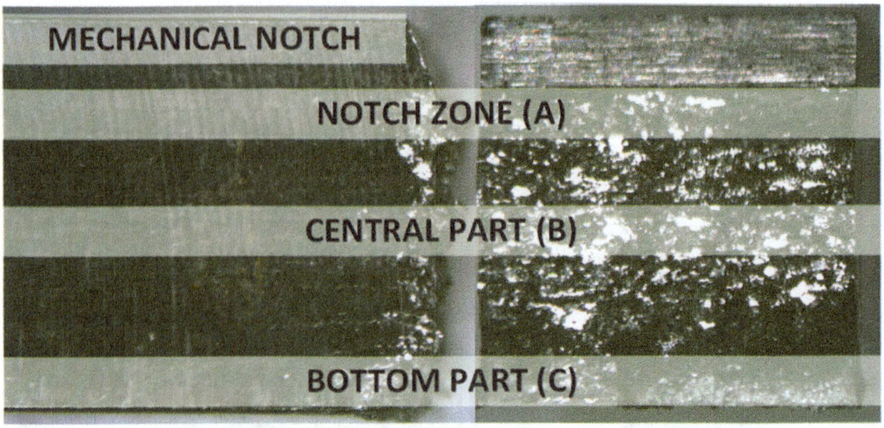

Fig. 1.5 Specimen after impact resistance test with marked characteristic zones

1.2.1 Ageing Processes in Low Carbon Constructional Steels

Ageing processes are influenced by among others, carbon and nitrogen content in the steel. Particularly fast ageing process existence is more likely for the metallic materials which contains less than 0.1% carbon content. It is considered that the ageing effect, caused by nitrogen, is greater than that caused by carbon reason. A characteristic feature of the low carbon, long-term operated mild steels is almost total absence of silicon (max. 0.07%). Compared to modern steels, the long-term operated metallic materials (from early 20th century) are more susceptible to zone segregation [4]. The results of detailed analyses are presented in the papers [16–20]. Based on papers [21, 22], three basic mechanisms can be distinguished steel ageing—microstructural degradation:

- Precipitation ageing mechanism—occurring after steel cools rapidly from a temperature at which the solubility of nitrogen and carbon is significant (about 590 °C) to room temperature. As a result of such ageing, nitrides and carbides are separated from the supersaturated solution;
- Ageing after excessive deformation—occurring under the influence of cold forming, generating an increase in dislocation. The dissolved nitrogen diffuses in the direction of dislocation, causing them to be blocked and, as a result, the brittleness of the steel;
- Dynamic ageing—occurs during hot forming. During this process, nitrogen during deformation can diffuse in the direction of dislocation. Due to the danger of the so-called blue brittleness, steel should not be processed. plastically in the temperature range from 200 to 350 °C.

Role of nitrogen in ageing processes has been presented many times in works [13, 23–27] but without reference to experimental results. They were taken into account

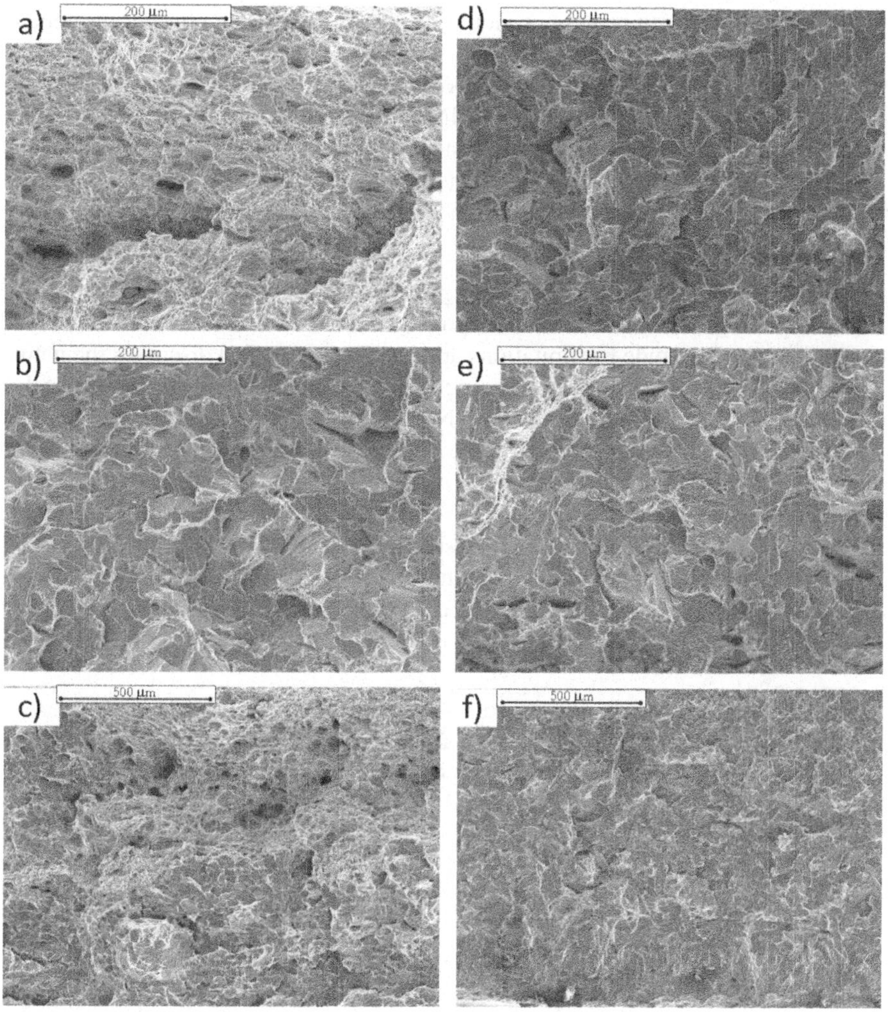

Fig. 1.6 Test fracture surfaces of the long term operated mild steel (1930) in the in-service condition **a** zone A, test temperature +20 °C, **b** zone B, test temperature +20 °C, **c** zone C, test temperature +20 °C, **d** zone A, test temperature −40 °C, **e** zone B, test temperature −40 °C, **f** zone C, test temperature −40 °C

only in the research cycle of Ishizaki and Imai [18–21, 23]. Here, will be presented only the most important conclusions obtained from this research:

- The transition temperature of plastic-brittle steel increases in the same proportion as the nitrogen content of the steel (Fig. 1.6);
- steels with a higher manganese content show a lower brittleness effect due to the increased nitrogen content;

- the effect of nitrogen is more pronounced in steels with high phosphorus content (Fig. 1.7);
- Al additive significantly reduces the effects of nitrogen secretion processes by favourable binding of nitrogen in AlN nitrides, therefore no reduction in the maximum impact strength values of steel is observed. The Al additive does not lead to a significant reduction in effects;
- increasing the ductile-brittle transition temperature caused by increased nitrogen content in steel [20].

The process of iron (i.e. puddle iron) ageing itself takes place in two stages. In the first stage, nitrogen and carbon atoms segregate around the dislocation, in the second stage, nitride and carbide precipitations appear. The paper [28] discusses the nature and type of nitrides emitted during ageing of low-carbon steels. It was observed that after five hours metastable nitrides ($Fe_{16}N_2$) appeared during ageing at the temperature of about +100 °C, while after five hours the first nitride separations (Fe_4N) appeared, and after more than 100 h the metastable phase completely disappeared. It should be emphasized that up to about +200 °C carbon was released from ferrite together with nitrogen as part of the metastable phase, as carbide nitride $Fe_{16}(N,C)_2$. These inaccuracies are justified by the authors [28], according to whom in the case of relatively low nitrogen saturation (about 100 ppm) only the separations occur of

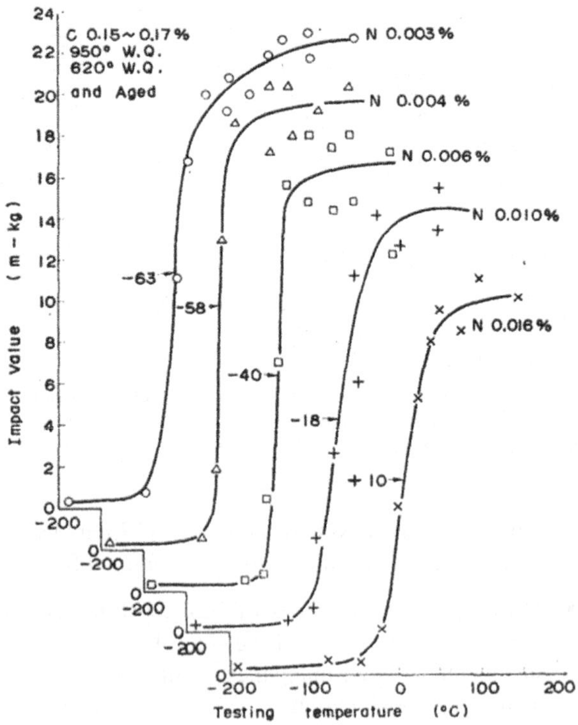

Fig. 1.7 Influence of nitrogen content on the transition temperature of plastic-brittle unalloyed steel (0.16% C, 0.25–0.35% Mn, 0.12–0.2% Si, 0.004% P, 0.02% S) after ageing [18]

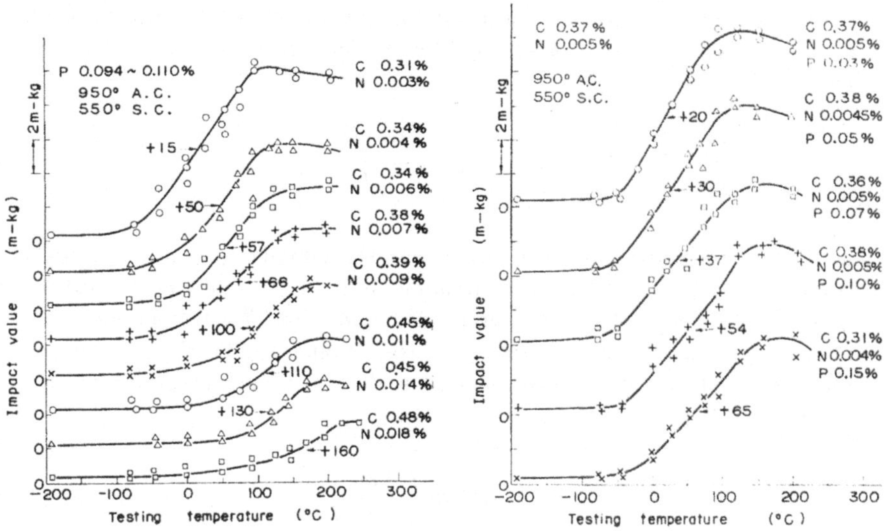

Fig. 1.8 Influence of nitrogen and phosphorus content on the impact strength of unalloyed steels [19]

the metastable phase ($Fe_{16}N_2$). For the nucleation of this phase, distant diffusion is required. The presence of dislocations is therefore necessary for such divisions. The effect is increased under the influence of variable stresses, causing an increase in the kinetics of separations. Diffusion in the presence of vacancies accelerates the ageing process of steel (Fig. 1.8).

Kocańda [29] presents and justifies the combined fatigue and ageing scheme of long-term operated steel in the range of:

- Limited fatigue strength—variable dislocation density in slip bands, increasing with the number of N cycles at constant stress amplitude caused the formation of saturated Cottrell atmospheres. This can lead to the nucleation of the new phases and their secondary disintegration, the complication of the dislocation blockade phenomenon, the formation of clusters of vacancies. As a result of creating clusters of vacancies and combining them into gaps of critical dimensions, microcracks are created;
- below fatigue strength—increasing gradually, at constant loading amplitude, the density of dislocation in individual slip bands is set at a certain level. Further activation and development of dislocation are accompanied by strong blockades, reinforced by the separations of carbides and nitrides. Stabilization of the slip bands is not disturbed by vacancies whose concentration in this load range is too low to create larger gaps.

The rapid diffusion of foreign dislocating atoms violates the "natural" ageing process under steady thermal and load conditions. In theoretical considerations concerning the kinetics of the development of precipitations phenomena justified on the dislocation ground, the Cottrell-Bilby model is adopted in accordance with [29] and [30]:

$$n(t) = n_0\left(\frac{ADt}{kT}\right)^{2/3} \tag{1.3}$$

where $n(t)$ represents the number of foreign atoms accumulating at time t around the dislocation, n_O—the number of foreign atoms in the initial material, D—diffusion coefficient, A—constant, k—Boltzmann constant, t—time, T—temperature.

The number of $n_v(t)$ atoms migrating to dislocation, related to the unit volume of the material, can be obtained by introducing to the equation (Eq. 1.3) the density of N_d dislocation [29]:

$$n_v(t) = \alpha N_d n_0\left(\frac{ADt}{kT}\right)^{2/3} \tag{1.4}$$

The concentration of atoms introduced into dislocation zones is equal:

$$c(t) = \frac{n_v(t)}{n_0} \tag{1.5}$$

It is assumed that n_O, N_d, A, D, kT are (in simplified terms) fixed. So:

$$c(t) = \left(\frac{t}{r}\right)^{2/3} = \frac{kT}{(\alpha N_d)^{3/2}AD} \tag{1.6}$$

The saturation of Cottrell's atmospheres will be achieved when one carbon atom falls on one dislocation line and on one atomic plane. The required time t can be determined, by inserting in the Cottrell-Bilby model (Eq. 1.3) the parameter of inter-atomic distance. The results of the calculations presented in the paper [29] show that for $T = 20$–200 °C and $n_O = 10^{20}$ at/cm^3 this time is in the range of 0.003–10 s. From the presented considerations it follows that ageing effects are possible both in the phase of nucleation and precipitation of with a relatively long loading time (cf. time to saturation of Cottrell's atmospheres). Kocanda [29] assumes that the precipitation mechanisms involved in deformation ageing overlap with the "natural" ageing process of steel. Apparent evidence of these separations in a periodic manner of loaded materials vary depending on the number of N cycles and load amplitude. According to the experimental and theoretical deliberations in [29], resulting from dislocation theories it should be noted that the material ageing, indicates a significant complexity of fatigue processes and their connection with "natural" and deformation ageing.

1.2.2 Strength Aspects of Degradation of Long-Term Operated Mild Low-Alloyed Steels

The problems discussed so far concerned mainly the degradation of microstructures. The impact of degradation processes on the strength properties of steel is not fully recognized—it depends on the material and operating environments. However, the papers [13, 18–21] have shown the consequences of changes in mechanical properties of microstructurally degraded materials. In general, it concerns on the:

- increase of the material hardness;
- increase in yield strength R_e (YTS) and tensile strength R_m (UTS);
- the decrease in ductility manifested mainly by the reduction of relative elongation in the static tensile test;
- the decrease in impact resistance and increase in temperature of ductile-brittle transition,

Material degradation manifests itself macroscopically in changes in mechanical characteristics caused by brittle phase separations—cementite Fe_3C_{III}. This results in a decrease in plastic properties, decrease in impact strength, increase in susceptibility to brittle cracking, which can be triggered more easily along grain boundaries than trans-crystalline. According to the concept of the theory of degradation of a multi-criteria system of information on the state of degradation of mechanical properties, the following—fundamental—types of studies are proposed:

- static tensile test;
- impact test and plastic-brittle transition temperature;
- Fracture toughness tests.

The static tensile test is one of the basic tests used to determine the properties of structural materials. Provides information on elongation and the probable degree of ductility loss. The impact of degradation processes better reflects the results of the impact tests. The characteristic feature of these tests in relation to the static tensile test is the ability to determine more precisely the changes in the strength properties of the material caused by long-term loads under structural degradation conditions. The purpose of impact tests is to determine the effect of dynamic loads on the conditions of the material's transition to a state of brittleness. Experiments show that metals at increased strain rate, as well as the previously mentioned and confirmed degradation phenomena, show a general tendency to increase yield stress and strength. The most common method for impact tests is an impact bending test, carried out on Charpy hammers, on notched specimens. The impact test (as well as the static tensile test) is carried out on the following materials in the original condition and after normalizing heat treatment procedure (in case of the lack of the virgin material state), which simulates the initial condition of the material before use. The results of impact tests depend on the temperature. In order to determine the plastic-brittle transition temperature, these tests are carried out for different ranges

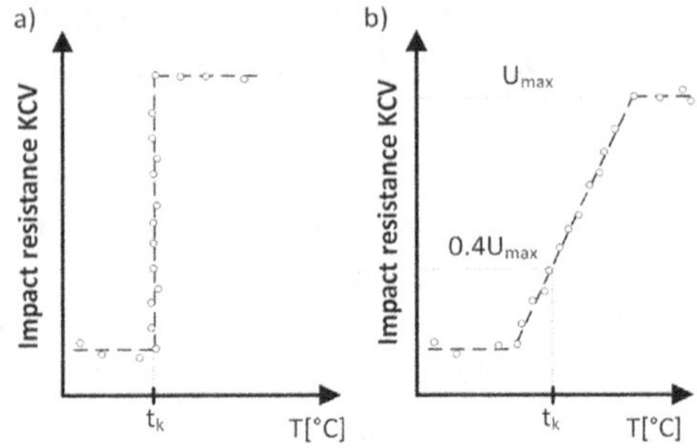

Fig. 1.9 Two different Charpy curves behaviour; **a** with distinct ductile-brittle temperature transition, **b** with ambiguous interpretation of the ductile-brittle temperature transition

temperatures. Most often, in the case of facilities operated in typical conditions (excavators, bridges, viaducts), the temperature range is from −40 to +20 °C. This temperature (t_k) is defined as the temperature at which at least 50% of brittle fracture surface is obtained during the impact test. The determination of the quantification of brittle/ductile fracture surface ratio takes place most often during fractographic analysis of fractographical fractions by scanning electron microscopy methods, as well as by means of quantitative methods of material engineering. Usually such studies are time-consuming and expensive. There are also approximate methods to determine this temperature resulting from the analysis of impact test results. Drawing measurement data from impact tests on a common consolidated diagram $U = f(t)$, in which U means impact strength expressed in J referred to the specimen cross-section area (in cm^2), allows obtaining one of the typical curves shown in Fig. 1.9. The transition to brittle state can be done in two ways: incremental (Fig. 1.9b) and distinct (Fig. 1.9a). In the paper [31] it is proposed to take as critical value the plastic-brittle transition temperature corresponding to 0.4 U_{max} temperature.

In practice, the criterion 35 J/cm^2 [32] is used to determine the plastic-brittle transition temperature. The ductile-brittle transition temperature is interpreted as a cut-off of the hypothetical function of the change in impact strength depending on the temperature for values KCV = 35 J/cm^2. The influence of microstructure degradation processes on the strength properties of steel was studied in the works [13, 33]. The aim of the performed experimental campaign was to compare the differences in the measured values of mechanical parameters resulting from the microstructural degradation changes. In accordance with the mentioned above assumptions, the research was conducted in two states, Post-operated (P) and normalized (N). According to the [13] all analyses were carried out on steel structural members of excavators marked as steel grade: S235JRG2, S235JR, S355JR (according to the old nomenclature: St3SX, St3S, 18G2, respectively). The results of static tensile testing, impact strength and resistance to pecan are shown there. In the case of a static tensile test (the results are presented in Table 1.1), it was shown that the differentiation of mean values of yield point: $R_{0.2}$ and R_m do not provide a clear answer to the assessment of the degree of degradation intensity on the measured yield/strength ratio—R_e/R_m (YTS/UTS).

Table 1.1 Mechanical properties of materials from excavator chassis

Designation of steel–state–no of operation years	YTS (MPa)	UTS (MPa)	A_5 (%)	KCV + 20 °C (J/cm^2)	KCV − 20 °C (J/cm^2)
1. S235JRG2–(St3SX)–P-55	294	452	33.0	15.8	5.7
2. S235JRG2–(St3SX)–N-55	299	564	21.7	58.7	9.9
3. S235JRG2–(St3SX[a])–P-55	405	575	22.0	29.5	19.5
4. S235JRG2–(St3SX[a])–N-55	394	564	25.5	32.0	20.5
5. S355JRG–(18G2)–P-25	424	611	28.0	95.0	71.0
6. S355JRG–(18G2)–N-25	382	601	31.0	118.0	81.0
7. S355JRG–(18G2)–P-55	405	611	30.2	69.6	34.2
8. S355JRG–(18G2)–N-55	378	559	27.4	95.7	31.9

[a]Increased Si content, P–post-operated state, N–normalized state

Impact strength values in the temperature range from −40 to +20 °C. In all cases, decreasing the temperature caused a reduction in impact strength. Normalization, each time resulted in an increase in impact strength.

The results of fracture toughness expressed by the J-integral and CTOD (Crack Tip Opening Displacement) tests (Table 1.2) were obtained for the samples in the normalized and post operated state at −20 and +20 °C.

Three-point bent SEN(B) specimens with a thickness of B = 10 mm were used for the study.

The results of these studies allow to formulate the following conclusions concerning steel S235JR (St3S)—segments A and B and S355JR (18G2)—segment C [13]:

- Critical integral J values for S235JR (St3S) (A and B) are very low, despite the relatively good strength properties (R_e and R_m). Normalizing does not significantly improve these values;

Table 1.2 Fracture toughness test of materials from excavator chassis

Designation of steel–state–excavator part (A, B, C)	J_{IC} +20 °C (N/mm)	J_{IC} −20 °C (N/mm)	CTOD +20 °C (mm)	CTOD −20 °C (mm)
1. S235JR–(St3S)–PA	15	13	0.013	0.009
2. S235JR–(St3S)–NA	20	16	0.032	0.026
3. S235JR–(St3S)–PA	22	19	0.045	0.032
4. S235JR–(St3S)–NA	29	25	0.054	0.048
5. S355JR–(18G2)–PC	92	71	0.16	0.11
6. S355JR–(18G2)–PN	108	89	0.19	0.15

- Structural members made of S355JR steel (18G2) are characterized by the good resistance to fracture described by critical J-integral values. Normalizing annealing increases only to a small extent of the J_{IC} value.

However, the study [13] did not determine further usefulness (or not) of resistance to cyclic loading in order to assess the degree of degradation. Special attention has been paid to the role of impact studies as a measure of the degree of degradation. The work [13] also has one more value—indicates the directions of searching for new indicators and measures of degradation. As a result, it can be mentioned that this indicator (impact resistance) allows finding relationship for the degradation of cyclic properties long-term operated bridge materials (discussed in next chapter). According to the dimensional analysis [34] Lesiuk and Szata [35] described the relationship for the cyclic strength loss for post-operated and normalized state.

References

1. Dudek D, Hawrylak H (1993) Zagadnienia teorii degradacji w długotrwale eksploatowanych maszynach roboczych., Problemy Maszyn Roboczych, nr 1, pp 129–141
2. Dudek D, Oziemski S, Sobczykiewicz W (1994) Elementy teorii degradacji maszyn, Problemy Maszyn Roboczych, N4, s 5–34
3. Dudek D (1996) Degradacja maszyn roboczych, teoria czy sztuka? Problemy Maszyn Roboczych 7:7–27
4. Dudek D, Pękalska L, Pękalski G (1996) Materiałowe aspekty degradacji maszyn podstawowych. Drugie Konwersatorium: Bezpieczeństwo oraz Degradacja Maszyn, strony. Wrocław-Szklarska Poręba, pp 37–49
5. Dudziński W, Pękalski G (2002) Materiałowe aspekty teorii degradacji. Polska Metalurgia w latach 1998-2002, tom II, strony. Komitet Metalurgii Polskiej Akademii. Nauk, Kraków, pp 488–494
6. Dudziński W, Haimann K (1997) Ocena stopnia degradacji materiałów stosowanych w maszynach górnictwa odkrywkowego z zastosowaniem metod mikroskopii elektronowej. Metody doświadczalne w budowie i eksploatacji maszyn, strony, Wrocław-Szklarska Poreba, pp 243–252
7. Dudek D, Oziemski S (2005) Interdyscyplinarny problem zarzadzania degradacja maszyn roboczych. Problemy Maszyn Roboczych 26:5–21
8. Dudek D (2004) Pytanie o nature zmian degradacyjnych w obiektach technicznych. Problemy Maszyn Roboczych 23:25–43
9. Dudek D (2002) Od prawdopodobieństwa do nieodwracalności procesów degradacyjnych w maszynach roboczych. Transport Przemysłowy (4(10)):57–60
10. Dudek D, Figiel A (2004) Problemy degradacji maszyn podstawowych górnictwa odkrywkowego. Górnictwo Odkrywkowe 2:38–40
11. Sobczyk K (2006) Zjawiska losowe. Panorama technika i matematyka, (3(7)):19–23
12. Dudziński W, Haimann K, Krajczyk A (1996) Zmiany mikrostruktury stali nieuspokojonych eksploatowanych w warunkach długotrwałych obciążeń zmiennych. Problemy Maszyn Roboczych 7:91–101
13. Pękalski G (2007) Procesy degradacyjne konstrukcji stalowych mostów Wrocławia z przełomu XIX i XXw. - na prawach rekopisu. Raport instytutowy, Instut Materiałoznawstwa i Mechaniki Technicznej, Wydział Mechaniczny, Politechnika Wrocławska, Wrocław
14. Haimann R (1974) Metaloznawstwo. Politechnika Wrocławska, Wrocław

15. Lesiuk G (2007) Strukturalne i wytrzymałościowe aspekty degradacji stali mostowych (na przykładzie mostu „Pomorskiego" we Wrocławiu). Praca magisterska, Politechnika Wrocławska, Wydział Mechaniczny, Wrocław
16. Rabiega J, Pękalski G (2007) Badania materiałowe mostów Pomorskich: Południowego, Środkowego i Północnego na rzece Odrze we Wrocławiu. Raport instytutowy SPR 9/2007,Politechnika Wrocławska,, Wrocław
17. Ferguson P (1985) Mechanical Properities of Nitrogen-Ferite. Metallurgical and Materials Transactions A 16:45–50
18. Imai Y, Ishizaki T (1950) Nitrogen as the alloying element in steels - I On the effect of nitrogen on the temper-brittleness in steels. Science reports of the Research Institute for iron, steel and other metals, Tohoku University
19. Imai Y, Ishizaki T (1965) Effect of nitrogen and other alloying elements on the low-temperature brittleness of steel. II: correlative effect of nitrogen and phosphorus. Science reports of the research institutes. Tohoku University. Ser. A, Physics, chemistry and metallurgy
20. Imai Y, Ishizaki T (1965) Effect of nitrogen and other alloying elements on the low-temperature brittleness of steel. III: Nitrogen fix and nitrogen quench-tempered steel. Science reports of the research institutes. Tohoku University. Ser. A, Physics, chemistry and metallurgy
21. Imai Y, Ishizaki T (1965) Effect of nitrogen and other alloying elements on the low-temperature brittleness of steel. IV: Effect of quench tempering. Science reports of the research institutes. Tohoku University. Ser. A, Physics, chemistry and metallurgy
22. W. Moron (red.). Migracja węgla i azotu w niskich temperaturach w stalach ferrytycznych i martenzytycznych. Wyd. Uniwersytet Śląski, Katowice, 1979
23. Bargel G, Anf Schulz HJ (2008) Werkstoffkunde. Springer Verlag, Berlin
24. Czapliński K (2009) Dawne wyroby ze stopów żelaza. Dolnośląskie Wydawnictwo Edukacyjne, Wrocław
25. Madaj A, Wołowicki W (2007) Budowa i utrzymywanie mostów. Wydawnictwa Komunikacji i Łączności, Warszawa
26. Ziółko J (1998) Ocena stali w diagnostyce konstrukcji stalowych. Warsztat pracy rzeczoznawcy budowlanego, Kielce. IV Konferencja Naukowo-Techniczna, 27-29 kwietnia
27. Helmerich R (2005) Alte Stähle und Stahlkonstruktionen. Materialuntersuchungen, Ermüdungsversuche an orginalen Brükenträgern und Messungen von 1990 bis 2003. Bundesanstalt für Materialforschung und -prüfung (BAM)
28. Skakov YuA (1962) All. Structural changes during aging of commercial iron. Metal Sci Heat Treat 4:105–108
29. Kocańda S (1962) O niektórych zjawiskach zmęczenia w stali o bardzo małej zawartości węgla. No 9/10 (121), (122) serii Dodatek do biuletynu WAT. Wojskowa Akademia Techniczna, Warszawa
30. Zhao JB, De AK, De Cooman BC (2000) Kinetics of cottrell atmosphere formation during strain aging of ultra-low carbon steels. Mater Lett 44:374–378
31. Szapownikow NA (1954) Badanie własności mechanicznych metali. Państwowe Wydawnictwo Naukowe PWN, Warszawa
32. Staub F (1979) Metaloznawstwo. Wydawnictwo "Śląsk", Katowice
33. Dudziński W, Pękalski G (2002) Materiałowe aspekty teorii degradacji. Polska Metalurgia w latach 1998-2002, number tom II, strony 488–494. Komitet Metalurgii Polskiej Akademii Nauk, Kraków
34. Ciavarella M, Paggi M, Carpinteri A (2008) One, no one, and one hundred thousand crack propagation laws: a generalized Barenblatt and Botvina dimensional analysis approach to fatigue crack growth. J Mech Phys Solids 56(12):3416–3432
35. Lesiuk G, Szata M (2014) Fatigue properties and fatigue crack growth in puddled steel with consideration of microstructural degradation processes after 100-years operating time. Proced Eng 74:64–67

Chapter 2
Mechanical Properties, Microstructure and Degradation Processes in Long-Term Operated Bridge Materials from the 19th Century and Early 20th Century

Abstract This chapter presents and discusses the typical properties of historical structural materials used in civil engineering. In particular, puddle iron and early 20th-century mild steel have been highlighted. The next part of this chapter presents the problem of structural degradation of long-term operated metallic materials (low carbon steel and puddle iron) and its potential impact of the degradation effect on the reduction of the materials' functional properties and, consequently, on the reliability of the steel structure. On the basis of case studies, a significant influence of structural degradation on the dynamic properties expressed in impact strength and cyclic properties was demonstrated. This chapter also presents the results of a low-cycle fatigue and fatigue crack growth test for historical materials from the Main Railway Station in Wrocław (Poland). The studies were conducted in the post-operated and normalized state as the one, which is supposed to imitate the initial state. The obtained results confirmed the hypothesis of the negative influence of degradation processes on the decreased fatigue life caused only by microstructural degeneration.

2.1 19th Century Puddle Iron and Early 20th-Century Mild Steel

Adopting new standardized regulations defining metallic materials, all low-carbon, 19th-century iron and iron-carbon alloys produced by the pudding method can be called nowadays *puddled steels*. Due to a historic reason, this ancient type of material will be called in this book "puddle iron"—as it was initially called in the 19th century. It should be taken into account that during the repair works of structures in civil engineering in the technical documentation of 19th-century steel structures all repairs or reinforcements of bridge structures were not always recorded. In the steel structures of mid-19th-century buildings, it is possible often to find members made from typical low-carbon constructional steel, although the whole structure of the object is made of the puddle iron. Distinguishing between the two steel grades without metallographic observations is not easy. It should be noted that at the end of the 19th century was dominated by puddle iron. However, the development of the metallurgy allowed for the gradual replacement of the puddle iron by new mild low-carbon steel. As a good

G. Lesiuk et al., *Degradation Theory of Long Term Operated Materials and Structures*, Structural Integrity 15, https://doi.org/10.1007/978-3-030-43710-7_2

illustration can be the data extracted from [1], referred to Great Britain: it can be estimated that in 1884 the whole production of puddle iron carried 55% and in 1885 only 49%, in Germany in 1889–54%, and in 1890–48%, while in 1913 only 1.3%. In Belgium, the production of puddle iron in 1895 was 55% and a year later 46%. The puddle iron was commonly used in 19th-century structural members, mainly for beams and ceiling joists, columns or load-bearing elements of roofs, as well as for concrete reinforcement. In the second half of the 19th-century puddle, iron began to be widely used in bridge building. However, chemical compositions and reference ranges of quantities of individual chemical elements are given in the literature with different accuracy. Based on work [1–5] chemical composition data were collected in Table 2.1 for puddle iron and old mile steel.

An essential property of structures made of puddle iron was both the large variety of their chemical composition, as well as the layered structure, probably resulting from small portions of the charge applied to the puddling furnace. A good illustration of this is the statistical distribution of the chemical elements in puddle iron obtained during Authors' expertise's (number of investigated objects >100) of puddle iron. In Figs. 2.1, 2.2, 2.3 and 2.4 are presented box-plots for C, Si, P, S respectively. In addition, the technological process itself caused significant contamination of such steel. Figure 2.5 shows the cross-section of a puddle iron specimen after a static tensile

Table 2.1 Typical chemical composition of the metallic materials existing in the late 19th and early 20th century (based on [1–5])

Material		C (%)	Si (%)	Mn (%)	S (%)	P (%)	N (ppm)
Puddle iron	Min.	0.0032	0.003	–	0.0034	0.011	37
	Max.	0.15	0.42	0.018	0.02	0.4	400
Old mild steel	Min.	0.026	0.026	0.036	0.063	0.009	110
	Max.	0.2	0.2	0.52	0.176	0.136	220

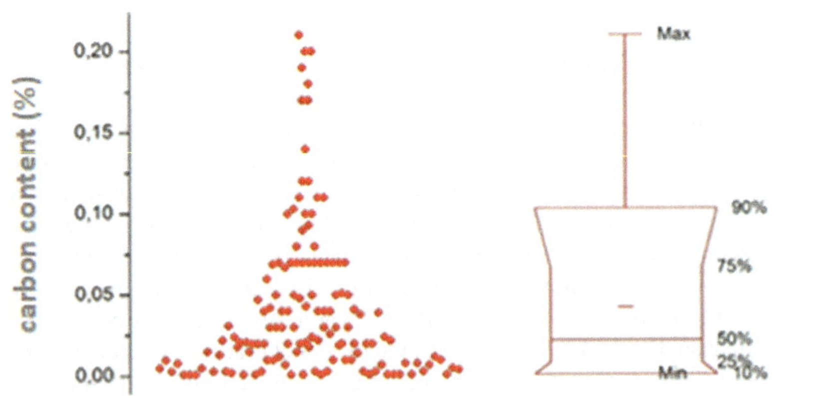

Fig. 2.1 Box-plot for carbon content in puddle iron, mean value %C = 0.042, min. 0.001%, max. 0.21%, n = 132, data source: [6]

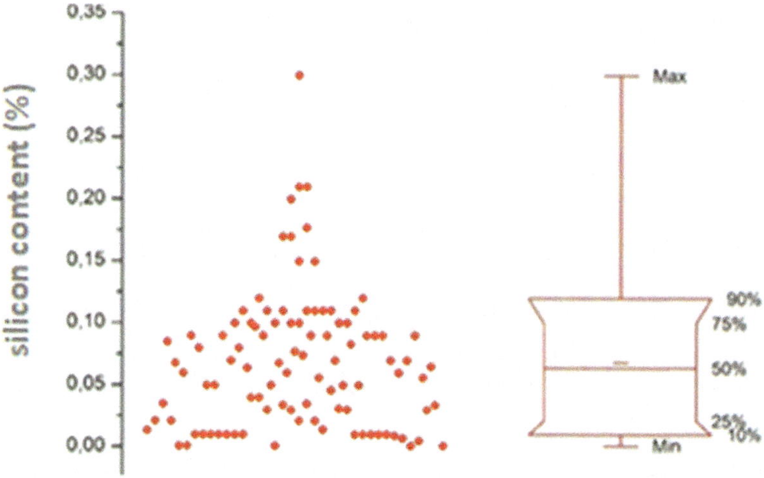

Fig. 2.2 Box-plot for silicon content in puddle iron, mean value %Si = 0.0684, min. 0.001%, max. 0.3%, n = 98, data source: [6]

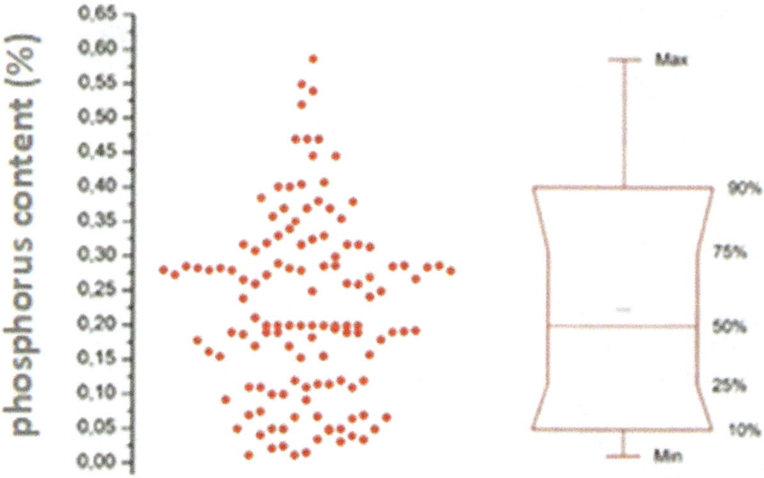

Fig. 2.3 Box-plot for phosphorus content in puddle iron, mean value %P = 0.224, min. 0.012%, max. 0.587%, n = 129, data source: [6]

test and the microstructure image typical for puddle iron. For comparison, Fig. 2.6 shows a picture of the macro- and microstructure of the old, early 20th-century old mild steel.

The method of production of puddle steel significantly influences its properties. The typical dimensions of a puddle furnace [8] length approx. 3.3–3.6 m at a width of 1.5–2.1 m wide (depending on the size of the furnace) and 1.5 m high. The walls of the furnace were made of bricks and had a thickness of about 23 cm, additionally

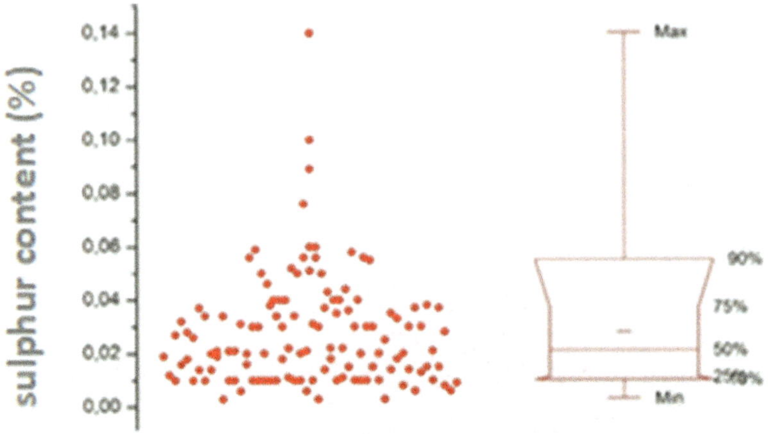

Fig. 2.4 Box-plot for sulphur content in puddle iron, mean value %S = 0.028, min. 0.003%, max. 0.14%, n = 123, data source: [6]

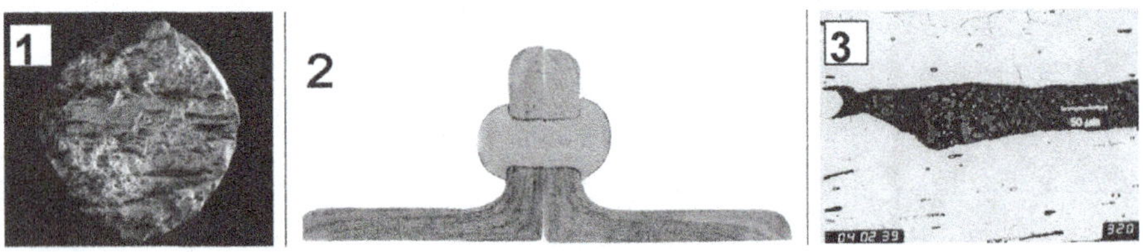

Fig. 2.2 Puddle iron: 1—fracture surface of the specimen from 1890 after a static tensile test, 2—the image of sulfur segregation (Baumann test) in a riveted joint (1882), 3—typical microstructure of puddle iron with slags [7]

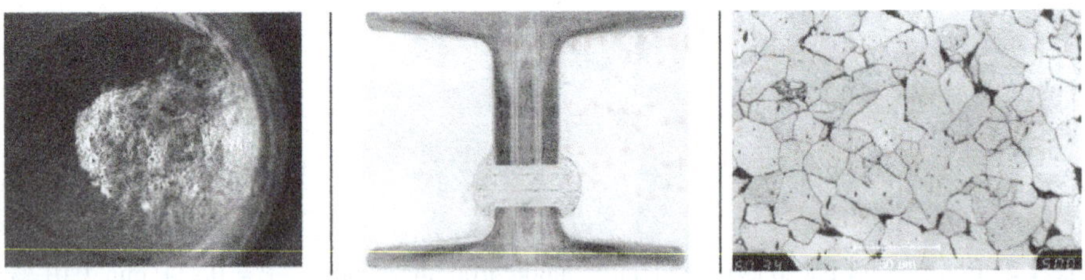

Fig. 2.6 Old, early 20th-century mild steel, 1—fracture surface of the specimen from 1890 after a static tensile test, 2—the image of sulfur segregation (Baumann test) in a riveted joint, 3—ferritic microstructure of steel with small areas of perlite [7]

were lined with cast iron plates. The chimney was usually 9–12 m high. The furnace was mostly elliptical in shape; 1.5–1.8 m long and 1–1.2 m wide. When the furnace was powered by white cast iron, the depth of the furnace did not exceed 50 cm, whereas when it was intended for the application of grey cast iron, the average depth was about 50–75 cm. Due to the formation of the significant temperature necessary for melting the charge, the grate had to be cooled continuously. There were also known varieties of more efficient, so-called double powder ovens, differing with an additional door pair, allowing for a double puddling process of steel at the same time. The efficiency of this furnace was twice as high as at the same time with a slight increase in the amount of fuel used to power one furnace [9, 10]. Figure 2.7 shows a sketch of a powder oven, used in Blain (Wales) with a water circuit. The components of this include [9, 10]; perforated cast-iron boxes (a), bridge (b), lower cast-iron plates (c), slab support bars (d), transverse supports for bars (e), side supports for supports (f), cast-iron plate (g), grate (h), hearth doors. (m), a set of pipes connected to the water basin (n). The fire is kept on the grate (h) passes through the bridge (b) to the hearth. By raising the door (m), the following takes place loading of the furnace. Mixing is carried out through individual blades—mixers through the hole (m). After loading the furnace, or more precisely its bed of pig iron in the amount of about 250–300 kg, it is brought to the boil. When the pig iron starts with a topic, the poodle (the worker operating the stove) closes the chimney. Then, as a result of the reduction of carbon and silicon, slag is formed on the surface of pig iron [2]. When the chimney is closed, continuous puddling begins (as shown in Fig. 2.8), i.e. mixing. It leads to a further reduction in carbon, silicon and manganese. As a result, the mass thickens, and mixing is no longer possible. To achieve a uniform decarburisation, the worker moves the parts below the agitator up and down. Then it tears them off, rotates, glues and rolls them into the pudding pan. The moulded metallic pieces are removed from the stove with special pliers (Fig. 2.8). They are then transported to the steam hammer in order to weld the loose mass and press out the remaining slag. The next and last stage is rolling [2]. In order to obtain 1,000 kg of sources, approximately 750–2000 kg of coal or coke must be used. The duration of the pudding process was on average about 2 h, so the daily output could not exceed 12–20 charges. In the 1860s, all three grades of metallic materials used in the 19th century, i.e. cast iron, puddle iron, mild steel were produced and used at the same time.

According to the [11], it is possible to determine the influence of chemical elements existed in steels on their properties:

- Manganese (Mn) is introduced into steel as a deoxidant. It reacts not only with oxygen (forming MnO manganese oxides), but also with sulphur, forming MnS manganese sulphides. Excess Mn is dissolved in ferrite. Due to the higher affinity of manganese to carbon, its concentration in cementite is three times higher than in ferrite. The manganese content in low-carbon steels does not exceed 0.8%; above this value manganese shall be considered as an alloying element. The influence of manganese on mechanical properties is manifested by ferrite hardening;
- The amount of silicon (Si) depends on the requirements of the steel deoxygenates and is about 0.05–0.35%. Like manganese, silicon dissolves in ferrite. The effect on mechanical properties is noticeable by increasing the hardness and strength of steel while reducing the ductility properties. Non-deoxidized steels contain traces of silicon, which makes them (like puddle iron) more plastic and more

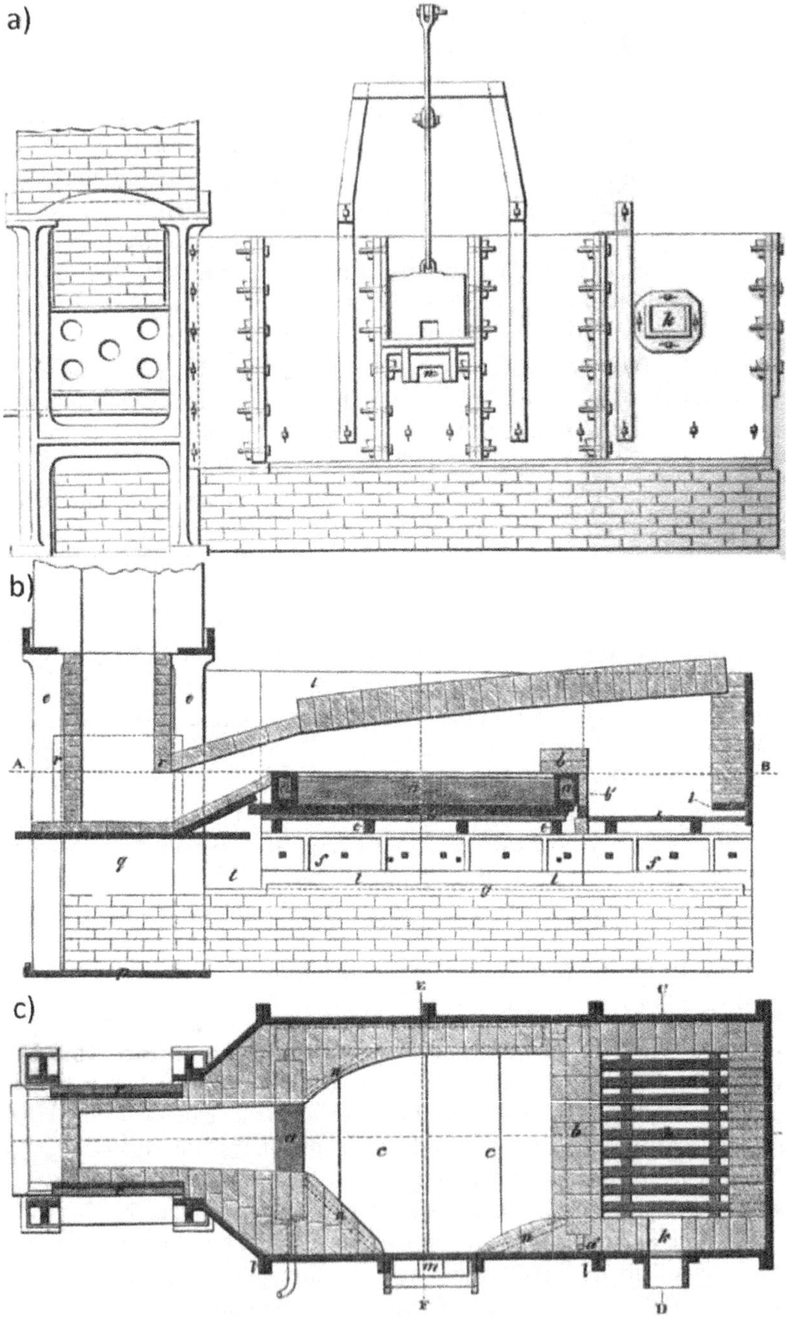

Fig. 2.7 Blain (Wales) puddle furnace: **a** furnace front, **b** cross section along the GH line, **c** horizontal section along the AB line [9]

Fig. 2.8 Iron puddler (left), ready iron pieces (right) [10]

susceptible to cold processing. Also, non-deoxidized steels are more inclined to ageing processes;

- Phosphorus (P) is a harmful admixture and cannot be removed completely. The presence of phosphorus is linked with the way of steel manufacturing. The effect of phosphorus on mechanical properties depends on the carbon content. In soft steels, such as puddle iron and old mild steel, phosphorus increases the yield strength and tensile strength. A particularly negative effect of phosphorus is noticeable in the impact strength—it causes an increase in the plastic-brittle transition temperature. In modern steels, the maximum permissible percentage of phosphorus in steel should not exceed 0.05%. The negative influence of phosphorus is particularly visible in "hard" steels, where phosphorus causes brittleness and susceptibility to separation cracks;
- Sulphur (S) forms FeS iron sulphide, which causes brittleness in the temperature range of 800–1000 °C due to low plasticity of the sulphides and above 1200 °C due to their melting. So-called hot brittleness is the main reason for considering sulphur as a harmful admixture. Sulphur bound in MnS manganese sulphides has the least harmful effect. Sulphur, or more precisely its segregation, is the cause of the failure of steel. This is due to the reaction of sulphides with hydrogen contained in the liquid weld metal, which results in the formation of H_2S.

In addition to the admixtures mentioned above, it is worth mentioning the role of gases such as nitrogen and oxygen, the presence of which causes ageing processes in steel. The solubility of nitrogen in ferrite is insignificant and at eutectoid temperature is about 0.1%. During cooling, it is relatively easy to saturate the ferrite, and ageing occurs at a low temperature. The nitrogen atoms accumulate near the defects of the crystalline structure in the form of dislocation, creating around them the so-called Cottrell atmosphere. Creation of Cottrell's atmospheres, as well as excellent nitride separations, blocked the dislocation movement. It deteriorates the plastic properties of steel while increasing brittleness and hardness. This phenomenon increases in soft steels after deformation, when the number of defects is relatively high [11]. Based on works [3, 11], the influence of the mentioned admixtures on the properties of steel is presented in Table 2.2. Slag inclusions and segregation in puddle iron during

Table 2.2 Influence of chemical elements existing in puddle iron and early mild on their properties

Property	C	Si	Mn	P	S	N
Tensile strength UTS	↑	↑	↑	↑	↓	↑
Elastic limit	↑	↑	↑	↑	•	•
Elongation	↓	↓	↓	↓	•	↓
Hardness	↑	↑	↑	↑	•	•
Impact resistance	↓	↓	↑	↓	↓	↓
Weldability	↓	↓	↑	↓	↓	↓
Hardenability	↑	↑	↑	↑	↑	•
Corrosion resistance	•	↑	•	↑	↓	•

rolling are arranged in plastically deformed chains. This band structure is characterized by high heterogeneity of chemical composition and significant anisotropy. It also determines the mechanical properties, both in the rolling and transverse direction. The tensile strength of puddle iron in the longitudinal direction is approximately 280–310 MPa at a total elongation of 7–25%. Young's E modulus of elasticity ranged between 170–200 GPa. Such a large variation in Young's modulus is significant, especially when converting deformations into stresses using linear analyses in numerical environments such as FEM (Finite Element Method).

According to [4, 11], it can be assumed that the following parameters characterized the mechanical properties of puddle iron manufactured in the second half of the 19th century:

- Density ρ: 7.6–7.8 (Mg/m^3);
- Elastic limit: $R_H = 150–160$ MPa;
- Yield strength: YTS = 210–290 MPa;
- Tensile strength: UTS = 300–400 MPa;
- Total elongation at break (in the rolling direction): $A_5 = 8–25\%$, Young modulus E = 170–200 GPa.

A good illustration of the typical yield stress and tensile strength data scatter for puddle iron is presented in Figs. 2.9 and 2.10 as a box-plot.

The low carbon content does not allow to obtain a martensitic structure puddle iron—it means that this type of material is not suitable for hardening. On the other hand, the early 20th old mild steel was characterized by the lack of slag inclusions and segregation of components and the high efficiency of the manufacturing process (Thomas-Martin steel). That is the reason why this material dominated in the early 20th century. Due to higher controlled manufacturing process, old mild steel was a basic structural material in the form of sheets, cylindrical sections and profiles. It was also in the form of forged steel, intended for further processing, such as bearing rollers, wedges, hinge pins, etc. In addition, the basic mechanical properties were better in comparison with puddle iron. As a comparison, it can be shown the distribution of the Charpy resistance—presented in Fig. 2.11.

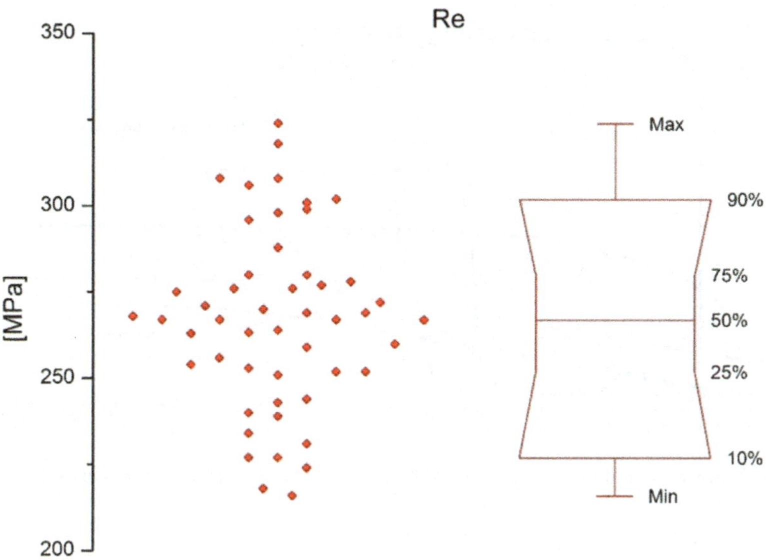

Fig. 2.9 Box-plot for yield stress (R_e) of puddle iron, min. 216 MPa, max. 324 MPa, mean value: 267 MPa, n = 49, data source: [6]

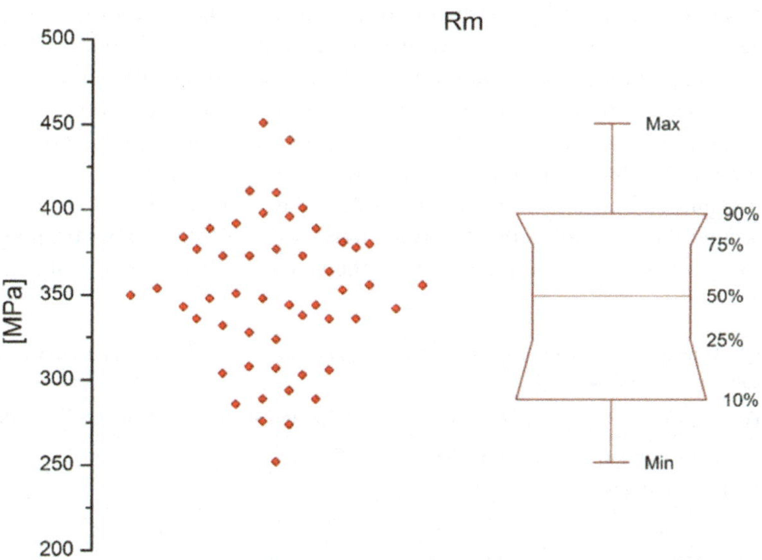

Fig. 2.10 Box-plot for tensile strength (R_m) of puddle iron, min. 252 MPa, max. 451 MPa, mean value: 350 MPa, n = 50, data source: [6]

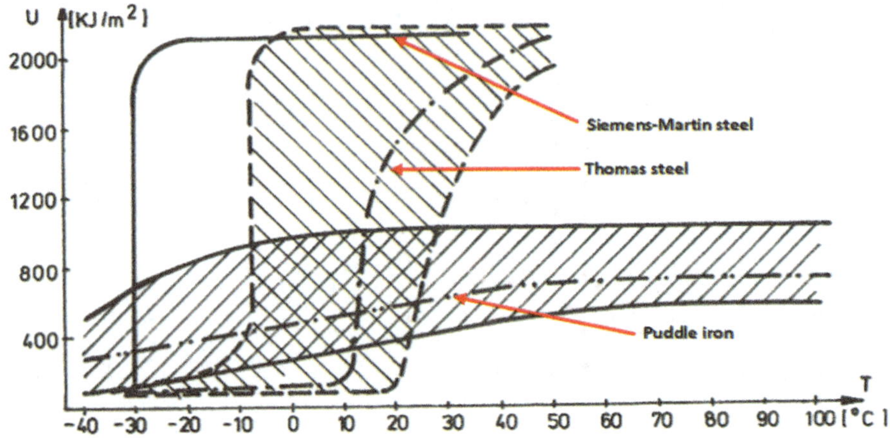

Fig. 2.11 Impact strength variation for puddle iron and old mild steel as a function of temperature; source data are redrawn based on [2, 4, 12, 13]

It can be concluded that puddle iron has a lower impact strength than the old mild steel. Exceptions may be values obtained in the negative temperature range. The impact strength of puddle iron is significantly less affected by temperature than old mild steel. The ductile-brittle temperature transition of the mentioned materials is above zero, which means that puddle iron and old mild steel is susceptible to cracking at ambient temperature. According to the literature, relatively little attention is paid to fatigue research on these constructional materials, with the exception of a paper [3] entirely devoted to this issue, as well as reported in Chap. 5 of this book.

In general, the basic mechanical properties of the 19th and early 20th century materials (puddle iron and mild steel) are collected in Table 2.3. From an engineering point of view, the allowable stress limits are important. Table 2.4 shows the maximum values of stresses in structural member for the years; 1920 and 1986 (Table 2.5).

Table 2.3 Typical mechanical properties of the puddle iron, old mild steel and modern S235J0 steel grade (based on [2, 4, 12, 13])

Property	Puddle iron	Old mild steel	Modern S235J0 steel
Young modulus E [GPa]	170–200	200–220	210
Shear modulus [GPa]	77	83	81
Elastic limit [MPa]	130–170	200–240	n/a
Yield point [MPa]	220–280	250–300	235
Tensile Strength [MPa]	330–400	340–450	360–510
Elongation A_5[%]	8–25	25–35	22–26
Compressive strength [MPa]	>220	>250	n/a
Shear strength [MPa]	>200	>320	n/a

Table 2.4 Allowable stresses in steel in puddle iron and old mild steel in the 1920 year—based on [14]

Strength/loading type	Puddle iron (MPa)	Old mild steel
Tension, compression, bending	108	140 MPa
Shear (rivets)	81	n/a
Shear	72	90 MPa
Shear (bolts)	63	n/a
Pressure strength	126	n/a

Table 2.5 Chemical composition of the tested materials; B and S

Material		C (%)	Si (%)	Mn (%)	S (%)	P (%)
"S"		0.06	0.17	0.1	0.028	0.290
"B"		0.03	0.03	0.13	0.025	0.198
Puddle iron	Min.	0.0032	0.003	–	0.0034	0.011
	Max.	0.15	0.42	0.018	0.02	0.4
Old mild steel	Min.	0.026	0.026	0.036	0.063	0.009
	Max.	0.2	0.2	0.52	0.176	0.136

The development of metal production technology in the second half of the 19th century (introduction of converters) has resulted in the production of steel with repeatable chemical compositions and less segregation of components. These steels were also characterized by a lower number of non-metallic inclusions in steps.

2.2 Description of the Microstructure Degradation Mechanism in Historic Bridge Steels

The structural aspects of the bridge steel degradation after one hundred and more years of service does not appear to be overdue. The degradation processes of the bridge steel structures (evidenced in [15–30]) do not take place as intensively as in brown coal excavators or power pipelines [31–35]. They are sometimes more intense only in the area of corrosive changes, which most often results from temporary and not fully thought-out repairs and renovations. The high durability of bridge structures results from the whole range of projects preceding their construction and operation. The analysis of the own data for the bridges located in Wrocław (Poland), the data on chemical compositions and basic strength properties of the tested mantellic materials leads to the conclusion that they are generally low-carbon, non-deoxidized steels. This means that they are more susceptible to degradation processes mentioned in Chap. 1. On the other hand, the analysis of the tensile test results of shows that the tested materials meet the standard requirements in bridge engineering. Therefore, if one approached the assessment of properties of these steels in a manner limited to the analysis of chemical composition and tensile testing, it could be concluded that these steels meet the requirements set for them. However, the degradation theory also takes into account the relationship between the different indicators in the post-operated and normalized state. Pękalski [15] concluded (on the basis of literature data and own research results) that the application of normalizing to degraded steels (950 C/heating time—2 h/air cooling) is a reconstruction (simulation) of structures and properties of the long term operated steels tested at the time of their application. This is a certain approximation because normalization usually causes fragmentation of ferrite grains and the associated increase in the yield strength of the material. On the other hand, during this heat treatment, carbides and nitrides, which are structural manifestations of degradation processes, are dissolved in austenite. Modern bridge steels are delivered in the normalized state. Also, after such heat treatment, the role of non-metallic inclusions in the post-operated and normalized state is not changed, although their influence on the strength properties and corrosion resistance is clear. Cooling of steel in technical conditions results in saturation of ferrite with carbon and nitrogen. Because of saturation with these elements at elevated temperatures, a state of thermodynamic imbalance is created, which is compensated by separation processes [36]. This results from the fact that the solubility of carbon and nitrogen at ambient temperatures in ferrite is lower than the content of these elements usually found in puddle iron and old mild steel. Therefore, during operation, long-term operated materials may be exposed to ageing processes at operating temperatures of bridge structures. The separation processes are diffusion processes, which in the case of interstitial atoms of metals also occur at room temperatures, and even at lower temperatures [37]. This results in the formation of various carbide and nitride precipitates in the steel structure. Therefore, nitrogen and carbon play a fundamental role in the degradation processes of microstructures of low-carbon old mild steel and puddle iron. However, in modern steels, standard DIN EN 10 025 restricts the nitrogen content up to 0.009%, but it is clear that in the 19th-century puddle iron and early 20th-century mild steel, this nitrogen regime could not be respected for technological reasons. In the regular, body-centred cubic (BCC) structures ferrite network, both carbon and nitrogen are located in the octahedral gaps of the crystalline network. The octahedral gap is larger than the tetrahedral gap and therefore if the dissolved atom has a diameter greater than 0.291d, it causes only two adjacent atoms in the network to move apart, rather than four as is the case with the tetrahedral gap. Therefore, the location of carbon and nitrogen atoms in octahedral gaps is easier in terms of energy and causes less deformation than in the case of the tetrahedral gap [38]. The mechanism of diffusion movements of N and C atoms is based on their jumping to adjacent octahedral gaps. At ambient temperatures it is supplemented by diffusion at grain and phase-to-phase boundaries. It is assumed that nitrogen at room temperature can diffuse to form G-P zones in planes {001} of BCC networks. As a result of increasing the temperature, these G-P zones are transformed into the a" ($Fe_{16}N_2$) phase, which then becomes an equilibrium g' (Fe_4N) phase [39]. On the other hand, in the work [40], it is reported that the metastable phase of Fe16N2 disappears after 5 h of heating at 100 °C and is replaced by the equilibrium phase of

Fe_4N. The processes of carbon and nitrogen release are often combined and result in the formation of a metastable $Fe_{16}(N,C)_2$ phase. The separation processes of nitrogen phases are accelerated in the presence of carbon with the occurrence of compression and tensile stresses [41]. In the analyzed group of historical materials from the nineteenth and early twentieth centuries, there is a microstructure of ferrite with tertiary cementite separations and sometimes few perlite colonies. Therefore, only cementite can be a "degradation phase" rich in carbon. Its separations may heterogeneously crystallize to cementite present at the borders of ferrite grains or form small separations in their area.

To summarize the separation processes occurring in the studied puddle iron and old mild steel, lead to an increase in their brittleness (increase in temperature of brittle-ductile transition) and to changes in their mechanical properties as a consequence of long-term operation under variable loads. The increased content of phosphorus (especially in case of the puddle iron), combined with the presence of nitrogen, can additionally increase the brittleness of such long-term operated material [42]. Structures made of structural steel sensitive to degradation processes should be particularly carefully maintained, especially as these processes can be strongly accelerated and intensified. This is the case in the heat-affected zones of welded joints and in larger zones in the case of the increased temperature operating conditions of e.g. fire. Therefore, welding spots should be areas of thorough microscopic examination in the case of e.g. riveted bridge structures where repairs have been carried out using welding technologies—as it is shown in Fig. 2.12.

Fig. 2.12 "Repaired" (by welding) riveted structure of the 19th century puddle iron from Wrocław, Poland

There are several ways of taking into account the degradation processes in steel structures. One of them is to determine the safe-time of operation. Published in [43], the standard [44] specifies the durability of bridge structures for 100 years. For brown coal mining machines, this time was defined as 30 years. This does not mean, that such structures should be withdrawn from use after the specified service lifetime. However, there is a presumption of the potential risk of their failure and strong need to investigate their technical condition in order to make decisions: repair, modernize, decommission and disassembly. The second method is the testing of the metallic material condition (specimens taken from the structure), which allows to determine the current strength properties and thus to take into account the long-term degradation process. This leads to the determination of the valid design strength criteria. Unfortunately, it is very rare that extensive strength tests can be carried out on specimens taken from e.g. the main bridge girders. More often, the possibilities of taking representative specimens are limited and the determination of the static properties e.g. R_m (UTS) and R_e (YTS) limits. In such cases, analyses of the chemical composition of steel and metallographic tests are particularly important, as they allow to determine the progress of microstructure degradation of the material and allow to predict its further changes as a function of the service life.

2.3 Microstructural Degradation and Its Influence on Mechanical Properties Based on the Case Study of the 19th Century Material from the Main Railway Station in Wrocław, Poland

In order to illustrate the potential impact of the degradation processes on the specific mechanical properties of the historical puddle iron, the results from the authors' study are presented below in the form of the case study. The materials for investigation was extracted from reconstructed Main Railway Station in Wrocław (Poland). During the renovation of this object, in the years 2010–2012, it was possible to obtain a significant amount of steel materials. It is worth to note that this material is equivalent to the 19th-century puddle iron commonly used in bridge engineering. In the book, the obtained materials were marked as S (rail profile) and B (I-section). All materials came from the basement ceilings of the hall Main Railway Station. According to the technical documentation, it is assumed that the materials date back to the second half of the 19th century. The material designated as steel B is the fragments of the I220 H-section, which is a reinforcement of the hall structure. In Fig. 2.13 is shown a view of the collected steel members and the location of the detailed points of specimens extraction.

After taking samples of steel elements, fragments of structures for metallographic tests were cut out after cleaning the materials from corrosion products. Samples were taken with cooling to prevent possible microstructural changes due to temperature. Five-elements (C, Mn, Si, S, P) analysis of chemical composition was carried out

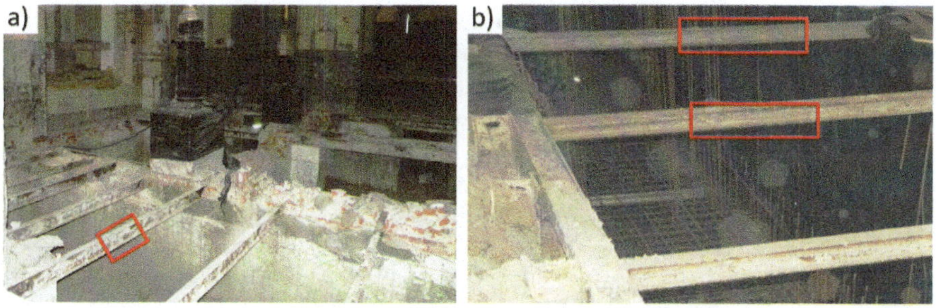

Fig. 2.13 Places of material extraction (B) and (S) from the Main Railway Station in Wrocław, Poland

by the gravimetric method. It is recommended to use this method due to the fact that chips are collected from different parts of the structural member. The results of chemical analysis are presented in Table 2.5.

The high phosphorus content is noticeable in the presented analyses. As it was mentioned in the previous section this is enough clear that the presented material belongs to the puddle iron group. However, this should be reflected in the micro-images of these materials. For observation of microstructures, light microscopy (LM) and scanning electron microscopy (SEM) was used. Observations of the metal structure were performed in etched (3% $HN0_3$) and non-etched state. In addition, for comparison two material states were considered; post-operated (marked as P) and after normalization (N). The microstructure obtained by LM (light microscopy) is shown in Figs. 2.14 and 2.15. Microscopic observations of the material marked as "S" in the non-etched state (Fig. 2.14a, d) shown a large amount of non-metallic inclusions in the form of chains, mainly multi-phase silicates. After etching the ferritic structure of this historical material has been identified. The low percentage content of carbon and silicon and a high content of phosphorus and sulphur (causing chemical composition segregation) have been confirmed during the microscopic examination (Fig. 2.14). The material exhibits also typical characteristics of symptoms of microstructural degradation processes. They are mainly manifested in the form of numerous brittle precipitations inside ferrite grains and thick envelope of the Fe_3C on the grain boundaries. The microstructure of "S" material is shown in Fig. 2.15 in the post-operated and normalized state. The microstructure of "S" material was shaped by ferrite grains and numerous non-metallic inclusions (A) with a continuous envelope Fe_3C_{III} on the grain boundaries of ferrite (B) and numerous microstructural degradation symptoms (C)—similarly like in material "B" case. For detailed analyses, SEM investigations were used—Figs. 2.16, 2.17, 2.18 and 2.19.

Microstructure observations ("B" material) in post operated state using SEM (Fig. 2.16a–d) have confirmed the presence of separations (C) inside ferrite grains (Fig. 2.16) and enabled for the identification of the types of phases in non-metallic inclusions. The results of the investigations were analyzed the energy spectrum of X-rays (the places of analysis were marked as two the arrows (marked by letter A) in Fig. 2.16 b. According to this spectrum—Fig. 2.16e–f, the predominant silicon

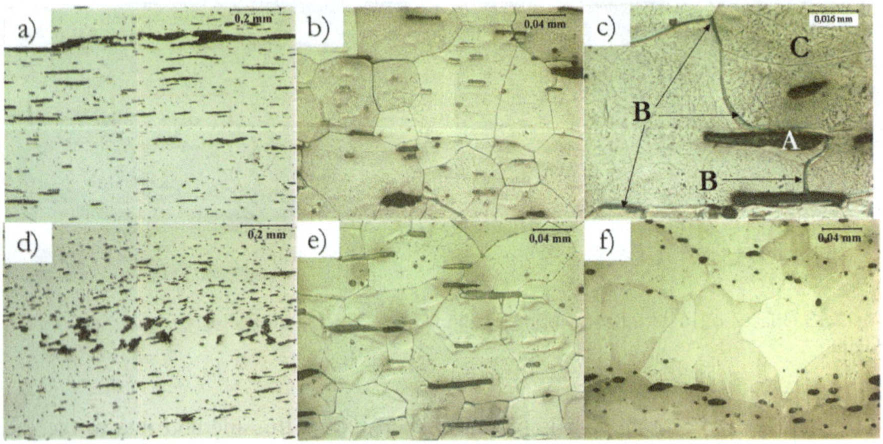

Fig. 2.14 Microstructure of puddle iron—marked as "B", **a** and **d** chains of plastically deformed non-metallic inclusions, **b** ferrite grain structure with continuous Fe_3C_{III} envelope on the ferrite grain boundaries (post-operated state—etched 3% HNO_3), **c** ferrite grains with numerous brittle precipitations (C) inside the grains with continuous cementite envelope on the grain boundaries (post-operated state—etched 3% HNO_3), **e** and **f** ferrite grains structure with numerous non-metallic inclusions (f) ferrite grains with multiple non-metallic inclusions (normalized state— etched 3% HNO_3)

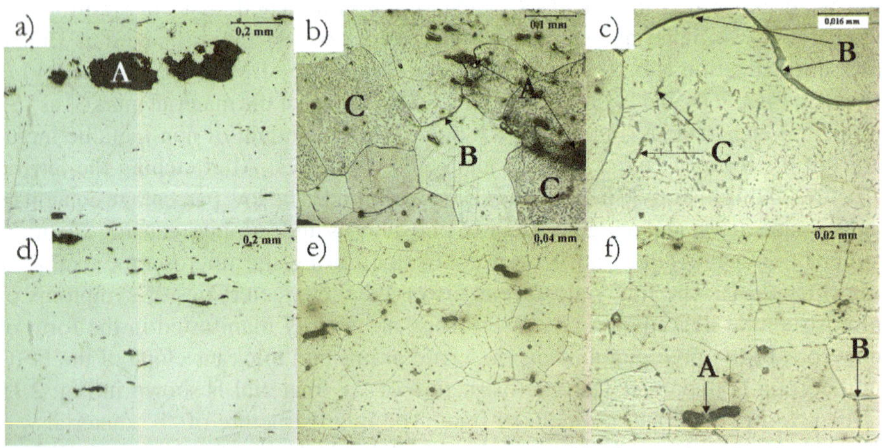

Fig. 2.15 Microstructure of puddle iron—marked as "S"; **a** and **d** Multiphase non-metallic inclusions (A), **b** Ferrite grains structure with multiple non-metallic inclusions (A) thick envelope of the Fe_3C_{III} on the grain boundaries (B) with numerous brittle phase precipitations (C)—(post-operated state—etched 3% HNO_3), **c** enlarged ferrite grain with a continuous $Fe_3C_{II}I$ envelope and numerous precipitations of the brittle phases (C)—(post-operated state—etched 3% HNO_3), **e** and **f** ferrite grains structure with non-metallic inclusions (A), noticeable is decreased number of brittle precipitation inside ferrite grains, (normalized state—etched 3% HNO_3)

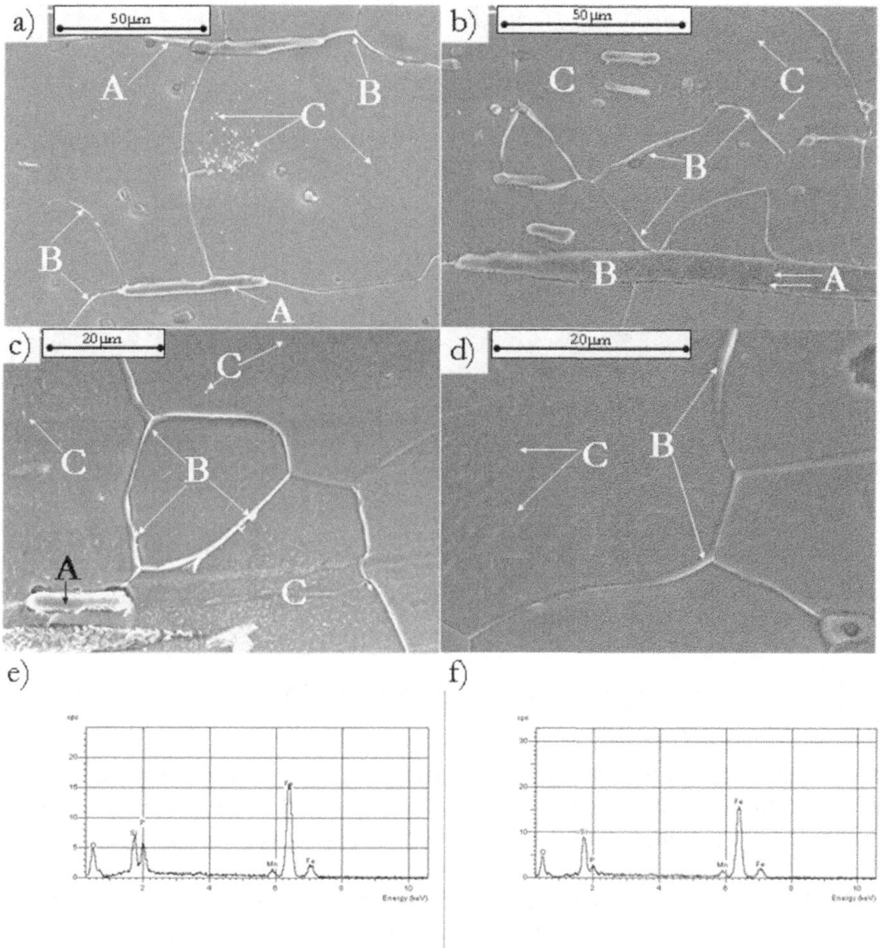

Fig. 2.16 Microstructure of the "B" material (obtained by SEM methods) in post-operated state

content was indicated, as well as the presence of phosphorus and oxygen. This confirms the multiphase structure of non-metallic inclusions in puddle iron.

The microstructure of the "B" material in the normalized state is shown in Fig. 2.17. This regenerating heat treatment procedure did not cause grains size reduction (compare Figs. 2.16 and 2.17). Obviously, as a result of normalization, the number, type and distribution of the non-metallic inclusions (A) has not been changed. However, the number and nature of brittle precipitations within the ferrite grains have been significantly changed. Ferrite grains with a reduced number of precipitations (C) is shown in Fig. 2.17a–c. These separations exhibited no tendency to be concentrated in specific areas (as opposed to the post-operated

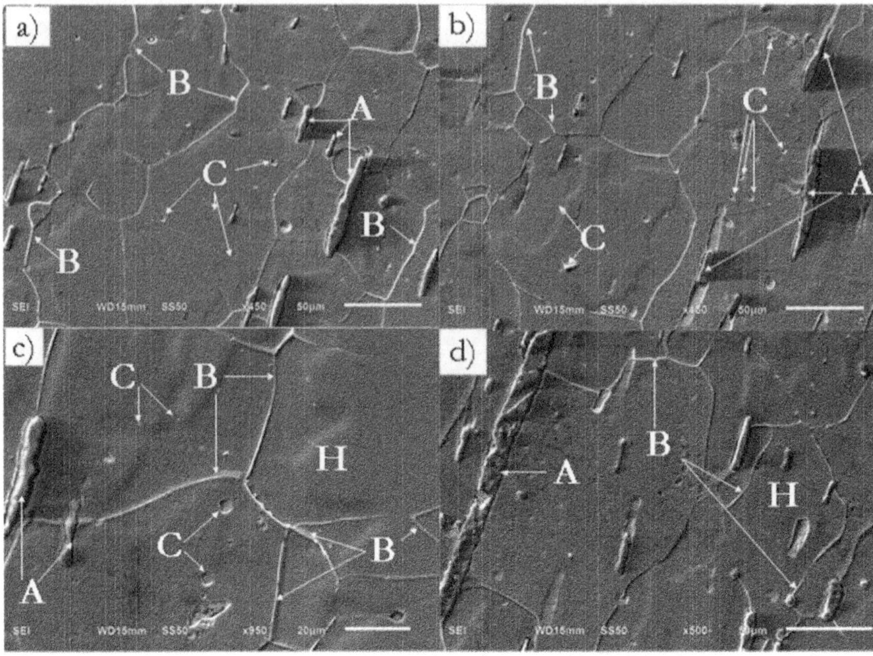

Fig. 2.17 The microstructure of the "B" material (obtained by SEM methods) in a normalized state

state—Fig. 2.16a). It is possible to also observe the ferrite grains free from degradation products, as determined by a marker (H) in Fig. 2.17c, d. Normalization has also resulted in a reduction of the continuous Fe_3C_{III} network around the ferrite (B) grains.

SEM images of the puddle iron marked as "S" are shown in Figs. 2.18 (post-operated state) and 2.19 (normalized state) respectively. The SEM observations in the post-operated state of the "S' have shown significant progress of degradation processes in the form of numerous brittle phase precipitations inside the ferrite grains (C)—Fig. 2.18a–d. In light of the evidenced progress of the microstructural degradation processes, it should be considered that the tested material "S" is close to the limit of service life. In order to identify the type of non-metallic inclusions in "S" material, the X-ray energy spectrum was analyzed (results in Fig. 2.18f), the location of the analyzed point is shown in Fig. 2.18b. The results of this analysis indicate that the non-metallic inclusions are multiphase inclusions with a predominantly silicate content with a relatively low content of oxygen and phosphorus compounds. Normalization (Fig. 2.19) caused significant changes in the microstructure of this material, manifested mainly in the form of a significant reduction in the number of microstructural degradation products inside the ferrite grains (C). After this heat treatment, it is also possible to observe degradation products free ferrite grains (H). However, in some cases, the remaining microstructural degradation processes grains are small needle-shaped precipitations (G).

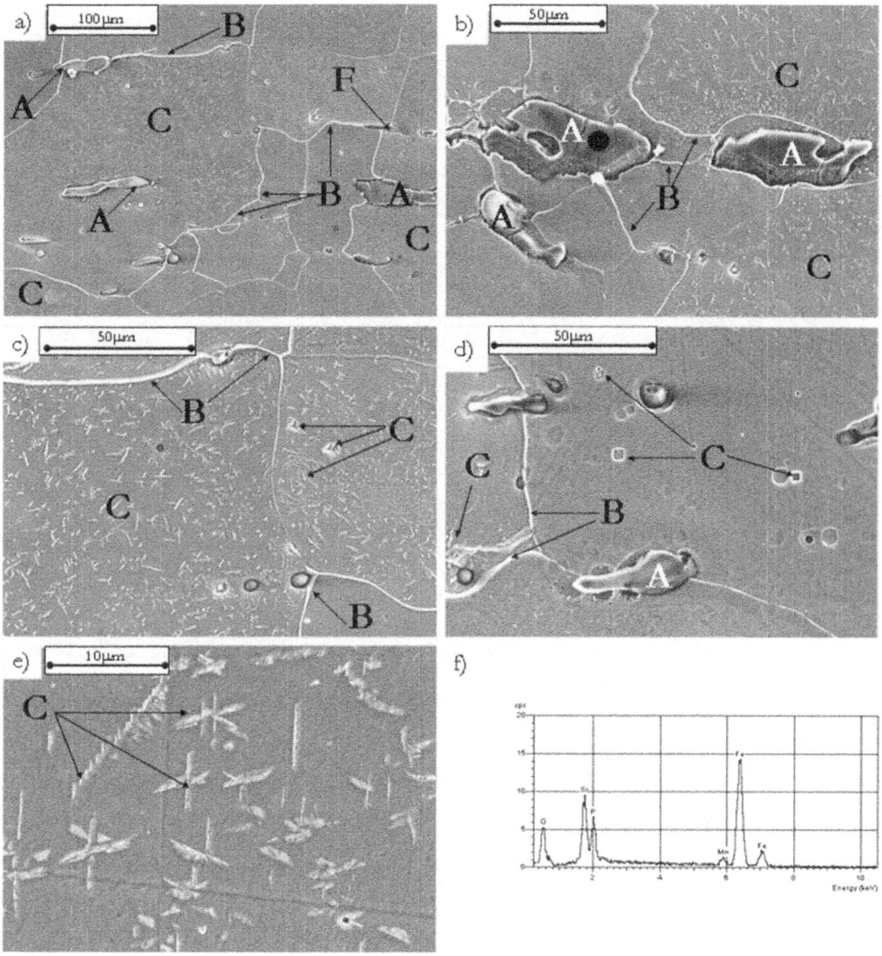

Fig. 2.18 Microstructure of the "S" material (obtained by SEM methods) in post-operated state

2.3.1 Charpy Test Results and Static Tensile Strength

As it has been already mentioned in the previous part of the book and according to Authors' research [15–25], the most sensitive degradation marker is impact strength estimated in Charpy test—comparing with the static tensile test. For the tested materials, impact resistance and static properties are collected in Table 2.6. Based on the data analyzed in the table, it can be noted that after normalization, the YTS/UTS ratio decreased each time, but in a significantly smaller than the impact strength (Charpy energy) in terms of KV_P/KV_N ratio. Therefore, similarly as in [15–25], it can be concluded that the tendency to brittle fracture expressed as a result of impact strength (Charpy energy) is an important parameter in the qualitative evaluation

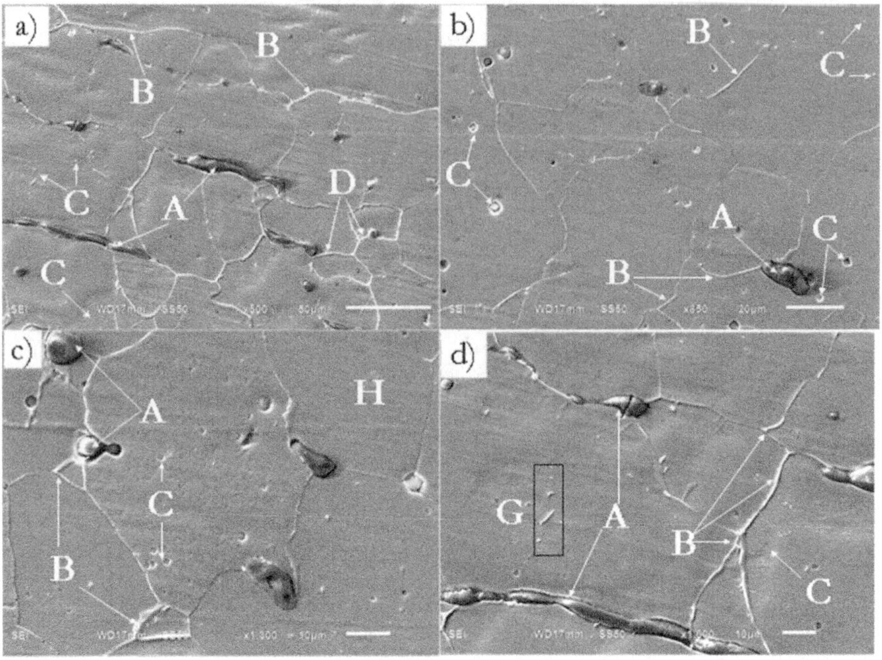

Fig. 2.19 The microstructure of the "S" material (obtained by SEM methods) in a normalized state

Table 2.6 Basic mechanical properties of the "B" and "S" puddle iron [17, 21]

Material/state	UTS (MPa)	YTS (MPa)	E (GPa)	A (%)	Z (%)	KV +20 °C (J)	KV –40 °C (J)	YTS /UTS
B/P	373	265	192	21	28	29	5	0.71
B/N	380	257	194	25	27	39	8	0.67
S/P	369	272	189	16	24	29	10	0.74
S/N	373	260	186	18	25	46	19	0.7
Puddle iron	330–400	220–280	170–200	n/a	n/a	n/a	n/a	n/a
Old mild steel	340–450	250–300	200–220	n/a	n/a	n/a	n/a	n/a

of material behaviour in the presence of degradation processes. The influence of microstructure degradation processes is reflected not only in the registered Charpy energy values but also in the fracture surface topography. For +20 °C, material "B" and "S" two representative specimens after the Charpy test were selected for microscopic analysis—Figs. 2.20, 2.21, 2.22 and 2.23. Fracture analysis was performed with stereoscopic microscopy methods (magnif. 8 × –10 ×) and scanning electron microscopy (SEM, magnif. 100 × –700 ×) for both materials. All observations were conducted in three characteristic zones of fracture surface; i.e. in the notch area

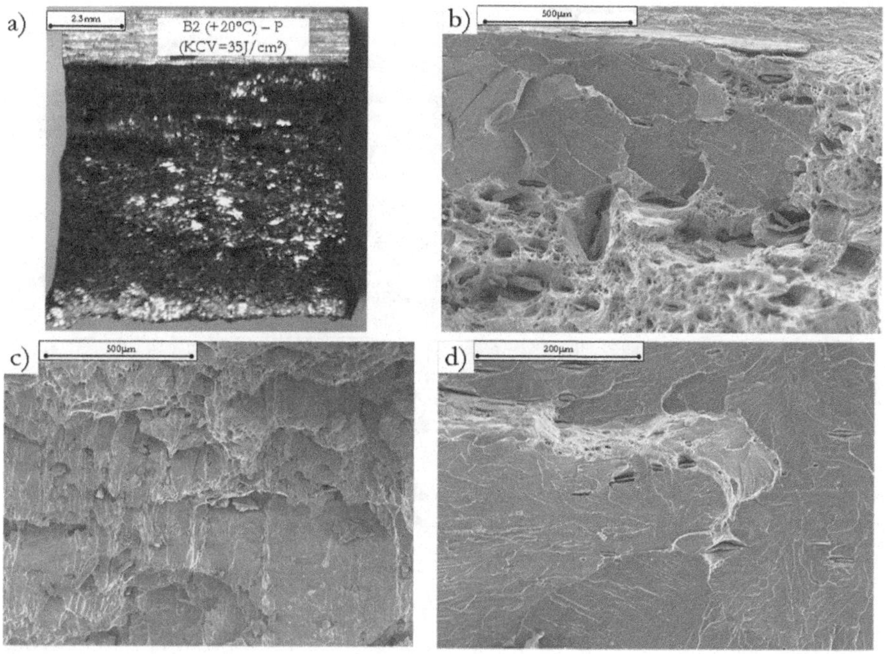

Fig. 2.20 The fracture surface of the specimen (material B, post-operated state, test temperature +20 °C, KCV = 35 J/cm^2) after Charpy impact test, **a** macro-view, **b** notch zone, **c** bottom part of the specimen, **d** central part of the specimen

(marked as zone A), in the central part of the specimen (marked as zone B) and on the bottom part of the broken specimen (zone C).

The fracture surface of the specimen (material B, post-operated state) after the Charpy impact resistance test (test temperature +20 °C) shown a mixed brittle-ductile character. In the zone near notch (A) area, there are large areas of brittle, transcrystaline fracture surfaces (Fig. 2.20b). Contrary to the post-operated state, in the normalized state ahead of the mechanical notch, dominates the plastic nature of fracture with typical features like; dimples with non-metallic inclusions—Fig. 2.21b. Also, the central part of the Charpy specimen is dominated by a plastic fracture type (normalized state). According to the SEM images, collected in Fig. 2.21d, contrary to the equivalent zone obtained for the specimen in the post-operated state (Fig. 2.20c—characterized by small areas of plastic fracture type cracking in between large facets of brittle transcrystaline fractures) in normalized state a noticeable is an increased amount of plastic-type of fracture areas, also mostly shaped by non-metallic inclusions pattern. However, in zone C of the specimens in both heat treatment states (P and N), a similar fracture character (a dominant shear type of fracture surface shaped by boundary effect and contact of the specimen surfaces) can be observed. However, in some part of the fracture surface located in (C) zone, due to fact of the

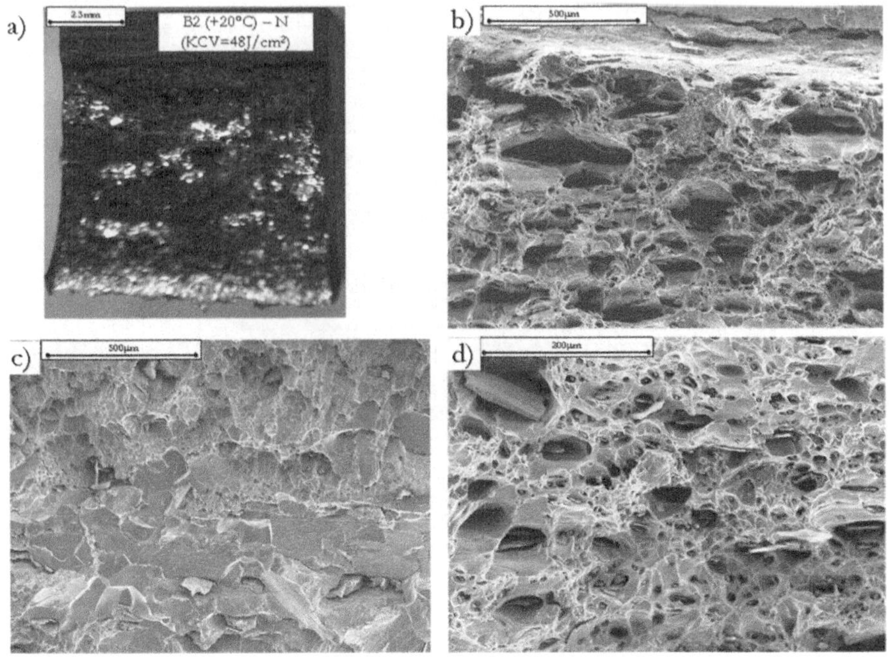

Fig. 2.21 The fracture surface of the specimen (material B, normalized state, test temperature +20 °C, KCV = 48 J/cm^2) after Charpy impact test, **a** macro-view, **b** notch zone, **c** bottom part of the specimen, **d** central part of the specimen

decreased material cohesion on the grain boundaries, it is possible to observe the fracture areas with symptoms of the intergranular type of fracture (Fig. 2.21c).

In the case of the material "S", the macroscopic features of the fracture surface in the post-operated state (Fig. 2.22) and normalized state (Fig. 2.23) differ significantly. In the post-operated state, the fracture surface is mostly shaped by brittle areas with local zones of plastic fracture. In the notch region (Fig. 2.22b), the cleavage type of fracture dominates. The brittle facets are separated by small areas of plastic fracture in the form of plastically deformed bridges. The fracture surface topography is also shaped by numbers of non-metallic inclusions. The central zone (B) of the specimen exhibits a smaller area of the plastic-type of fracture existing only in the form of small plastically strained parts of the materials, mainly separating facets of brittle transgranular fractures (Fig. 2.22d). However, in normalized state (Fig. 2.23) the fracture mechanism in near notch area (A) and in the central zone (B) is dominated by plastic-type of fracture with small areas of brittle, transgranular type. Also, macroscopic view on the specimens (compare Figs. 2.22a and 2.23a) reflects higher ductility of the specimen after normalizing including higher plastic deformation of the specimen cross-section. In the zone (C), as in case of material "B" (in both states), the similar fracture surface is observed. The different natures of fracture type for both materials are reflected in higher values of the Charpy energy—Table 2.6.

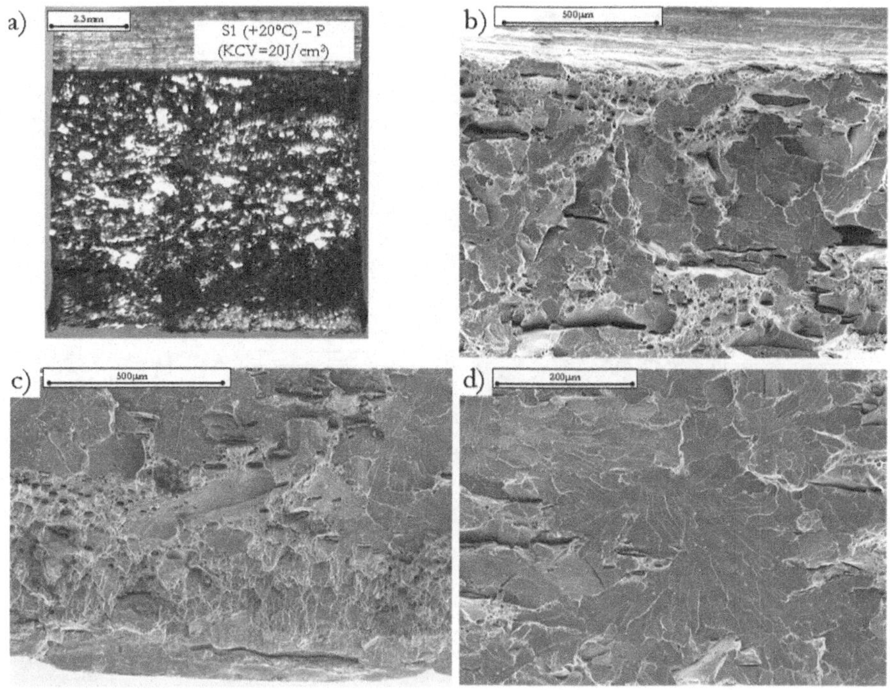

Fig. 2.22 Fracture surface of the specimen (material S, post-operated state, test temperature +20 °C, KCV = 20 J/cm^2) after Charpy impact test, **a** macroview, **b** notch zone, **c** bottom part of the specimen, **d** central part of the specimen, [17]

2.4 Influence of the Degradation Processes on the Fatigue and Fatigue Fracture Properties of the Long-Term Operated Puddle Iron in a Post-operated and Normalized State

As mentioned in the previous sections of the book, it is not always possible to extract a sufficient number of specimens for testing in the case of long-term operating objects. Fatigue tests, in particular, require this. Therefore, puddle iron material data related to fatigue are poorly described in the literature, and if so, never comprehensively. In the studied case, the number of material collected from the "S" beam enabled to perform a complete series of tests of LCF (low cycle fatigue) and HCF (high cycle fatigue) and FCGR (fatigue crack growth rate, also for material B).

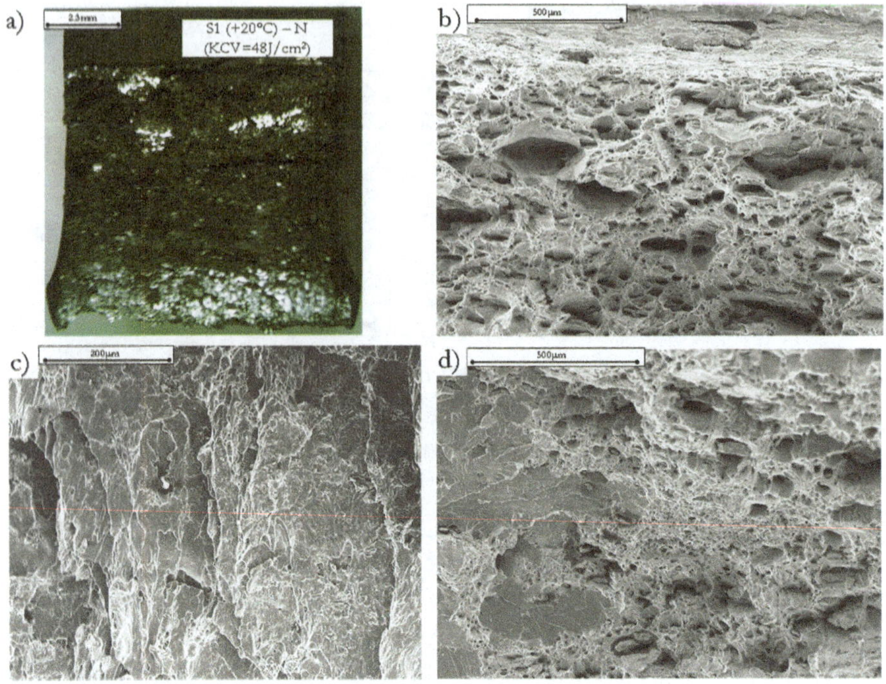

Fig. 2.23 The fracture surface of the specimen (material S, normalized state, test temperature +20 °C, KCV = 48 J/cm^2) after Charpy impact test, **a** macro-view, **b** notch zone, **c** bottom part of the specimen, **d** central part of the specimen, [17]

2.4.1 Impact of the Degradation Processes on Low Cycle Fatigue Material Behaviour

All tests were carried out in accordance with the recommendations of the Polish [45] and American [46] standards on cylindrical specimens. The shape and dimensions of the samples are shown in Fig. 2.24. The tests were carried out on the MTS809 testing machine using a 50 kN force load ce ll. The MTS extensometer (model: 632.27F-30) was also used for the experiment—strain range $\Delta\varepsilon = \pm\ 2\%$. The frequency of loading was f = 0.2 Hz, with the maintained strain ratio $R\varepsilon = -1$. The tests were programmed at four levels of total strain $\Delta\varepsilon = 0.4, 0.5, 0.6, 0.7, 0.8\%$. Six samples were tested at each level, and the criterion of test interruption was the complete separation of the specimen into two parts in accordance with the recommendations of the standards [45, 46]. During the study, the following signals were recorded: total strain, force F, displacement of the specimen and temperature using thermovision camera. The test stand is shown in Fig. 2.25. During loading, the increase of the specimens' temperature, over the number of total reversals to failure—$2N_f$, did not exceed 10 °C.

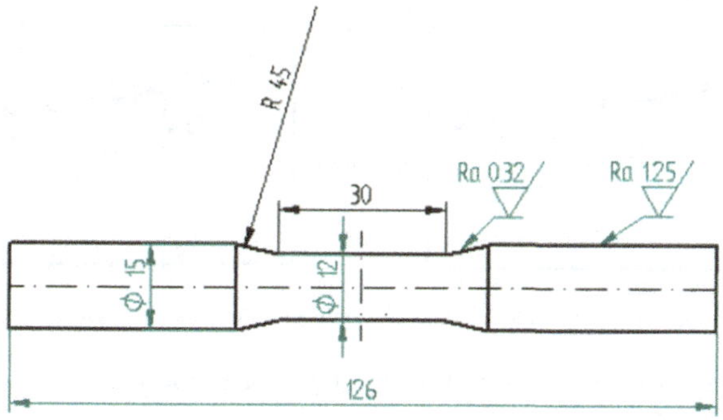

Fig. 2.24 Scheme of the specimen designed for LCF test

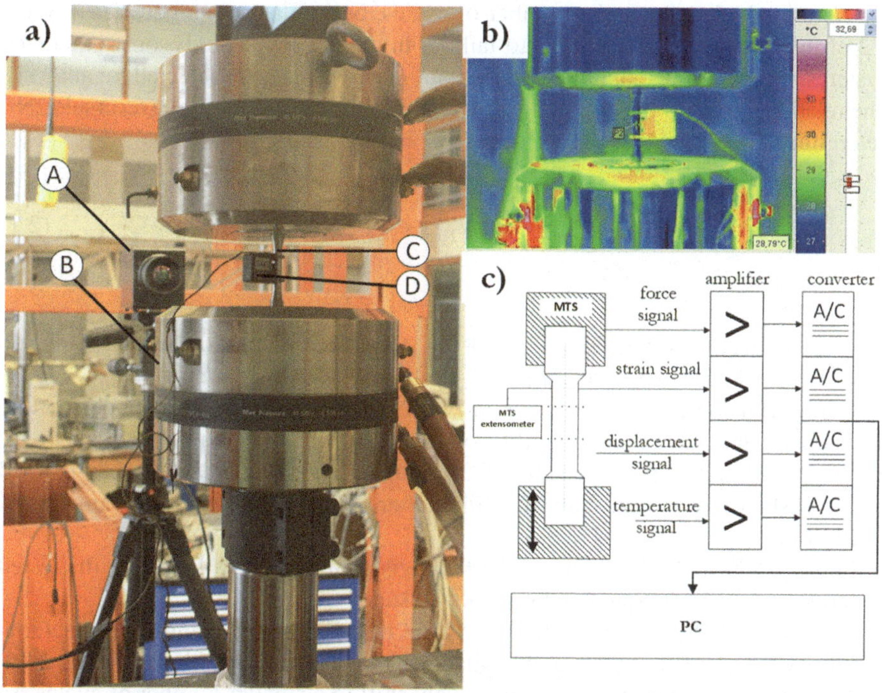

Fig. 2.25 Low cycle fatigue test measurement stand: **a** specimen during the test - A—thermovision camera, B—MTS 809, C—specimen, D—MTS extensometer, **b** registered thermogram, **c** scheme of the measuring system

Fatigue testing in the LCF regime is particularly important from the point of view of material behaviour in the range of elastic-plastic deformation. After the initial stage of cycling, associated with the cyclic deformation of the material (hardening, softening, etc.) there is a period of stabilization, called the saturation state. During

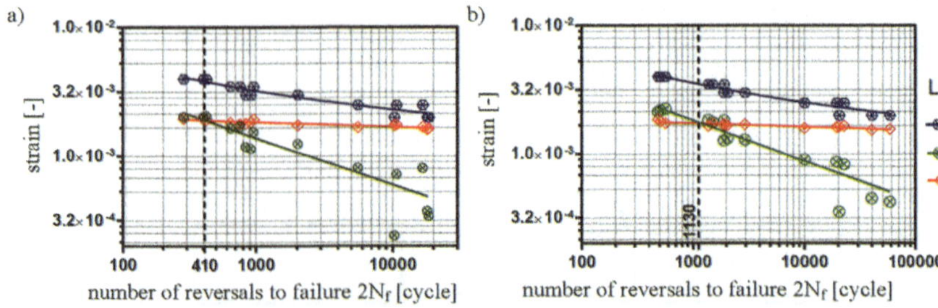

Fig. 2.26 Coffin-Manson curve for puddle iron ("S"): **a** in the post-operated state, **b** in the normalized state [21]

this stage, the hysteresis loops are no longer subjected to further changes. This state is determined after a specific number of cycles, but no later than between 1/3 and 1/2 of the total number of cycles to failure. In the presented study, only stabilized hysteresis loops, corresponding to the half fatigue lifetime were selected for analysis. The relationship between the total strain and the number of to failure (reversals $2N_f$) determines the law proposed by Coffin [47] and Manson [48]:

$$\frac{\Delta\varepsilon_c}{2} = \varepsilon_f^{'}(2N_f)^c + \frac{\sigma_f^{'}}{2}(2N_f)^b \tag{2.1}$$

where:

$\Delta\varepsilon_c$—total strain range ($\Delta\varepsilon_t = \Delta\varepsilon_s + \Delta\varepsilon_{pl}$; $\Delta\varepsilon_s$ elastic strain range, $\Delta\varepsilon_{pl}$—plastic strain range), b, c—exponents, σ'_f, ε'_f—cyclic stress, strain parameters.

Coffin-Manson curves are shown in Fig. 2.26. As it is noticeable, in case of the normalized state, the cyclic-ductility of such material is much better than in the post-operated state. It also results in $2N_t$—transition number of cycles—marked by the dashed line in Fig. 2.26. In Table 2.7 are collected all cyclic parameters of Coffin-Manson law (Eq. 2.1).

Figures 2.27 and 2.28 demonstrate the registered hysteresis loops for the tested material in its post-operated and normalized state.

In general, it should be noted that in LCF tests, with worsening of the material ductility (due to microstructural degradation), the intersection point $2N_t$ (Fig. 2.26) of the "$0.5\Delta\varepsilon_{pl}$" and "$0.5\Delta\varepsilon_s$" progressively shift to a smaller number of cycles An interesting feature of the low strain amplitude loading was the disappearance of a yield point in the post-operated state (Fig. 2.27a). After normalization in the first load

Table 2.7 Cyclic parameters of the Basquin and Coffin law for puddle iron ("S") [21]

Material/state	ε'_f (−)	c (−)	R^2	σ'_f (MPa)	b (−)	R^2
S/P	0.01699	−0.3632	0.90	442	−0.03651	0.78
S/N	0.01626	−0.3165	0.92	378	−0.02702	0.74

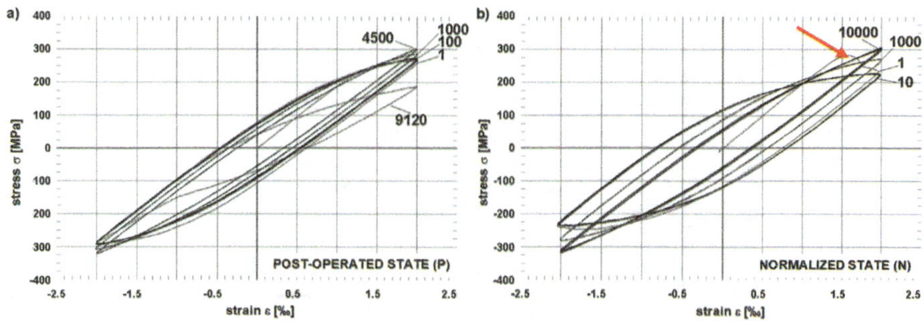

Fig. 2.27 Representative hysteresis loops for puddle iron "S" (total strain $\Delta\varepsilon = 0.4\%$): **a** in post-operated state ($2N_f = 18260$ cycles), **b** in normalized state ($2N_f = 40104$ cycles)

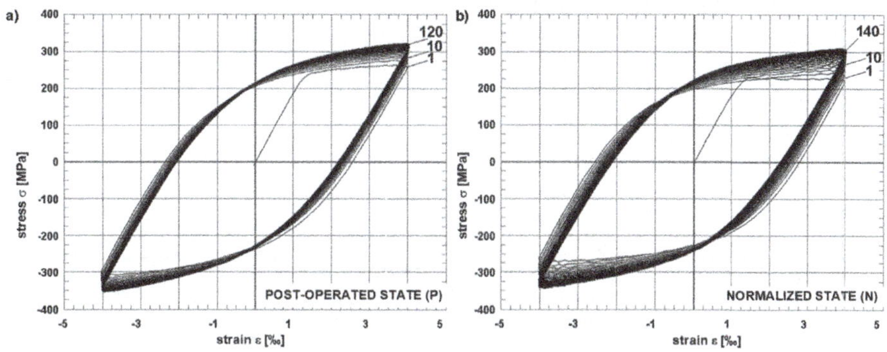

Fig. 2.28 Representative hysteresis loops for puddle iron "S" (total strain $\Delta\varepsilon = 0.8\%$): **a** in post-operated state ($2N_f = 424$ cycles), **b** in normalized state ($2N_f = 586$ cycles)

cycle, it was possible to observe a well-marked yield point (indicated by an arrow in Fig. 2.27b). After drawing stabilized tips (Fig. 2.29) of the hysteresis loops, the cyclic stress-plastic strain curves were constructed for "S" puddle iron material in the post-operated and normalized state.

2.4.2 Impact of the Degradation Processes on High Cycle Fatigue Material Behaviour

In accordance with standards [49, 50], the high cycle fatigue (HCF) tests were carried out for the puddle iron "S" in two states, i.e. after the operation and normalized one. The shape and dimensions of the HCF specimen are shown in Fig. 2.30. The aim of the tests was to determine the S-N curves for comparison of the cyclic behavior "S" puddle iron into two states (P and N). The tests (fully reversal $R = -1$) were carried out at controlled stress levels corresponding to $\Delta\sigma/2$ =167, 185, 208, 242,

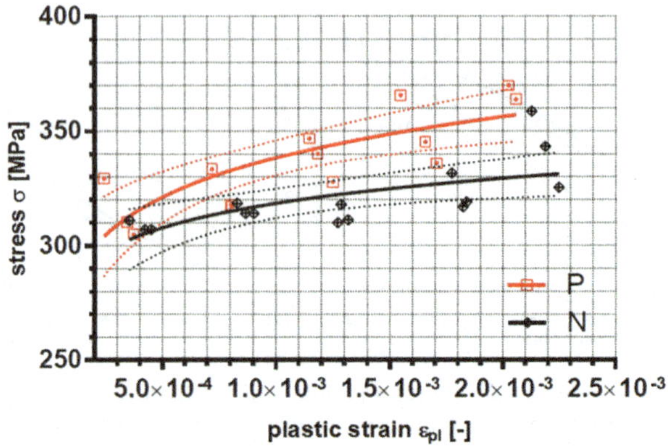

Fig. 2.29 Cyclic stress-plastic strain curve for puddle iron “S”

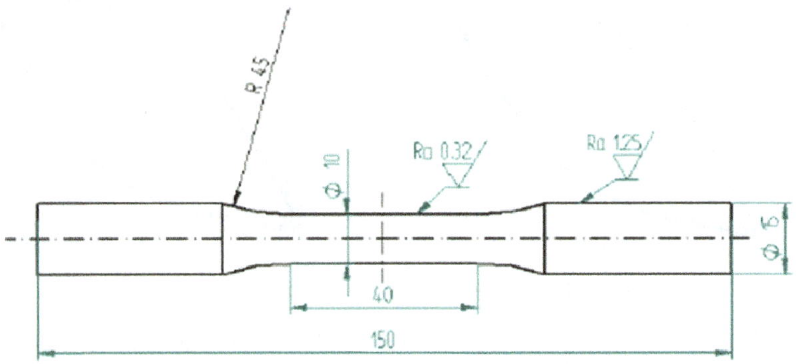

Fig. 2.30 Shape and dimensions of the test specimen designed for HCF test (all dimensions in mm)

280 *MPa* with load frequency f = 15 Hz. At each load level 6 samples were subjected to cycling loading and the number of cycles to specimen failure was recorded. The same measurement stand (Fig. 2.25) as in the LCF tests was used for HCF experimental tests. The temperature of the specimens was also periodically monitored using thermovision camera. During the tests, the temperature of the specimen, according to the recommendations of the standard [47], did not exceed 50 °C. As a result, the S-N—curves for both material states are shown in Fig. 2.31.

Analysis of the S-N curves (Fig. 2.31) allows concluding that fatigue life of the specimens in the normalized state is significantly higher (95% confidence interval marked with dots) than in the post-operated state.

Fig. 2.31 S-N curve for puddle iron "S" in post operated and normalized state

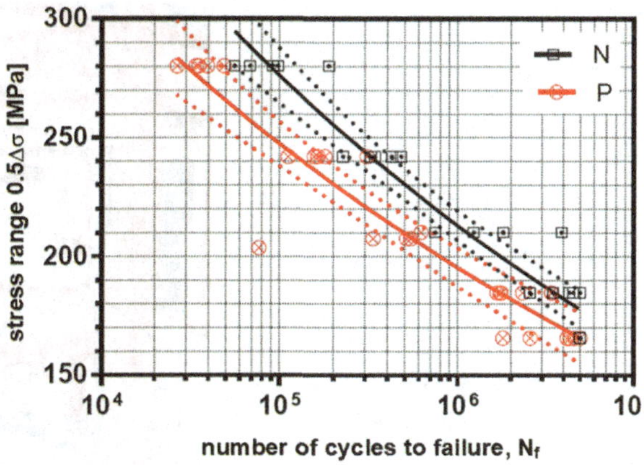

2.4.3 *Impact of the Degradation Processes on the Fatigue Crack Growth Resistance in Puddle Iron*

Fatigue crack growth rate (FCGR) experiments were conducted for all types of materials, according to standard [51], using compact tension specimens (CT). The shape and dimensions of the specimen are shown in Fig. 2.32. The FCGR measurement stand is shown in Fig. 2.33. All experiments were conducted using a constant load amplitude method and ΔK decreasing test type in order to characterise the near-threshold regime of the FCGR curve. Fatigue crack growth rate curves were calculated using the secant method. The load frequency f was 12.5 Hz with a constant stress ratio R equal to 0.1. Fatigue crack length was monitored during the experiment

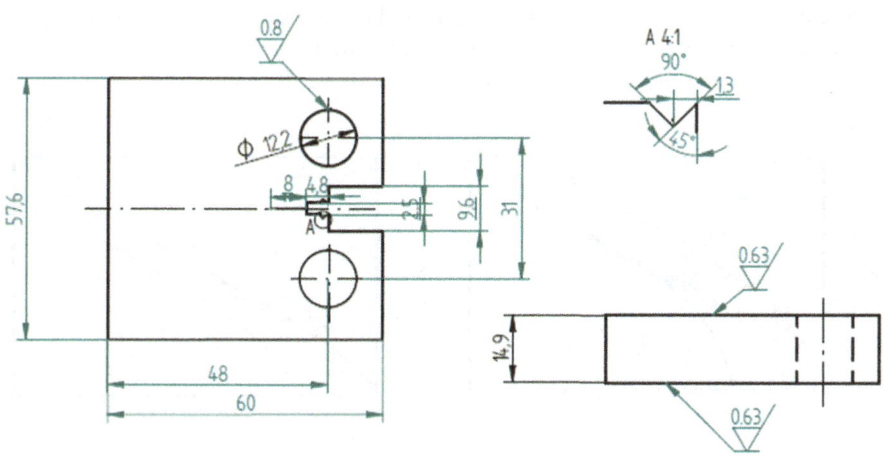

Fig. 2.32 Shape and dimensions of the test specimen designed for FCGR test (all dimensions in mm)

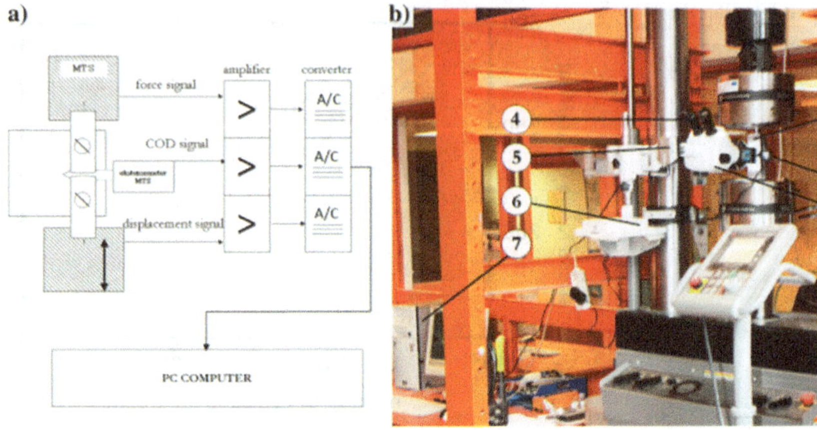

Fig. 2.33 Measurement test stand for fatigue crack growth rate; **a** scheme, **b** real measuring system; 1—Compact Tension specimen gripping system; 2—Extensometer, 3—Stereoscopic microscope, 4—Eyepieces, 5—Digital camera, 6—Tripod, 7—PC

using a compliance method [51]. Fatigue crack length was periodically corrected using direct measurements by stereoscopic microscopy. Mechanical notches were prepared using the electro-discharging machine with a wire diameter of 0.3 mm. Before the proper tests, the specimens were appropriately prepared i.e. surfaces of the specimens were polished to a mirror-like surface, necessary for of observing the fatigue crack lengths. All specimens were pre-cracked according to the ASTM E647 [51] requirements. During this phase, the final ΔK value did not exceed 15–17 MPa·m$^{0.5}$. All experiments were carried out in the same way for materials in two states, i.e. post-operating and normalised.

According to Fig. 2.34, the noticeable is higher data scatter for puddle iron than for modern mild steels. It is also noticeable that the fatigue crack growth rates for post-operated states are higher than for normalised one. It can be concluded that the

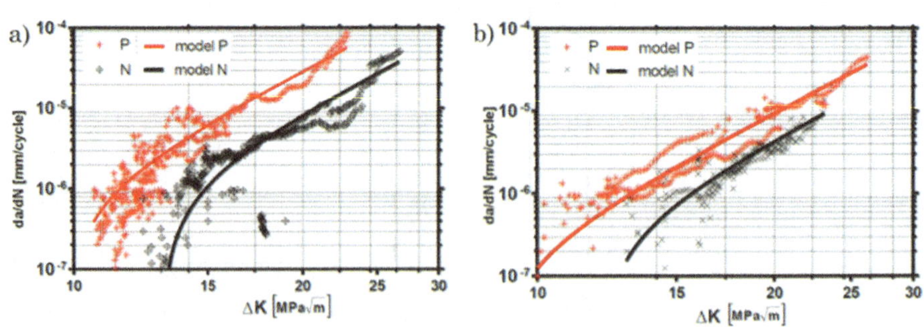

Fig. 2.33 Fatigue crack growth rate diagrams for "S" puddle iron (left) and "B" puddle iron

degradation processes reflect its impact stronger in the near-threshold region of the KFFDs (Kinetic Fatigue Fracture Diagrams).

References

1. Helmerich R (2005) Alte Stähle und Stahlkonstruktionen. Materialuntersuchungen, Ermüdungsversuche an orginalen Brükenträgern und Messungen von 1990 bis 2003., Bundesanstalt für Materialforschung und -prüfung (BAM). Berlin
2. Czapliński K (2009) Dawne wyroby ze stopów żelaza. Dolnośląskie Wydawnictwo Edukacyjne, Wrocław
3. Cremona C, Patron A, Johansson B, Larsson T, Eichler B, Höhler S (2007) Improved assessment methods for static and fatigue resistance of old steel railway bridges. Project co-founded by the European commission within the sixth framework programme (2002–2006)
4. Madaj A, Wołowicki W (2007) Budowa i utrzymywanie mostów. Wydawnictwa Komunikacji i Łączności, Warszawa
5. Helmerich R, Kühn B, Nussbaumer A (2007) Assessment of existing steel structures. A guideline for estimation of the remaining fatigue life. Struct Infrastruct Eng 3(3):245–255
6. Natusiewicz K (2015) Strukturalne procesy degradacyjne stali zgrzewnej, Praca dyplomowa, Promotor doc. dr inż. G. Pękalski, Wydział Mechaniczny, Politechnika Wrocławska, Wrocław
7. Łabecki H (1841) Górnictwo w Polsce. Opis kopalnictwa i hutnictwa polskiego pod względem technicznym, historyczno-statystycznym i prawnym - tom I. Drukarnia Juljana Kaczanowskiego, Warszawa
8. Overman F (1854) The manufacture of iron, in all its various branches. T.K. and P.G. Collins, Printers, Philadelphia
9. Percy J (1864) Metallurgy-iron and steel. John Murray, Albemarle Street, Londyn
10. Belford P (2004) Extra special best best: black country iron puddling and wrought iron manufacture in the nineteenth century. Historical Metallurgy 38:47–59
11. Haimann R (1974) Metaloznawstwo. Politechnika Wrocławska, Wrocław
12. Ziółko J (1991) Utrzymanie i modernizacja konstrukcji stalowych. Arkady, Warszawa
13. Kozar R (1970) Badania materiałów, elementów i konstrukcji. Budownictwo betonowe, Arkady
14. Rozporządzenie Ministerstwa Robót Publicznych nr 6342-10-IV-213, z dnia 4 marca 1920r
15. Pękalski G (1996) Material aspects in the degradation theory–conception of multi–criterion system of estimation of the state of material, Systems, vol 3, s 108–129
16. Konat Ł, Pękalski G, Rabiega J (2006) Degradation processes in bridge steels of the turn of 19–th century, 6th International symposium of Steel Bridges, Prague, s 11
17. Lesiuk G, Szata M, Bocian M (2015) The mechanical properties and the microstructural degradation effect in an old low carbon steels after 100-years operating time. Arch Civ Mech Eng 15(4):786–797
18. Lesiuk G, Szata M (2011) Aspects of structural degradation in steels of old bridges by means of fatigue crack propagation. Mater Sci 47(1):82
19. Lesiuk G, Szata M (2010) Degradacja mikrostruktur elementów konstrukcyjnych pochodzących z mostów wzniesionych na przełomie XIX i XX wieku. Zeszyty Naukowe/Wyższa Szkoła Oficerska Wojsk Lądowych im. gen. T. Kościuszki, pp 96–109
20. Krechkovska H, Student O, Lesiuk G, Correia J (2018) Features of the microstructural and mechanical degradation of long term operated mild steel. Int J Struct Integr 9(3):296–306
21. Lesiuk G, Szata M (2014) Fatigue properties and fatigue crack growth in puddled steel with consideration of microstructural degradation processes after 100-years operating time. Proced Eng 74:64–67
22. Lesiuk G, Rabiega J, Szata M (2011) Problem kruchości stali zgrzewnych w warunkach strukturalnej degradacji-badania statyczne, cykliczne i dynamiczne. Zeszyty Naukowe/Wyższa Szkoła Oficerska Wojsk Lądowych im. gen. T. Kościuszki, pp 254–269

23. Lesiuk G, Rabiega J, Szata M (2013) Mikrostrukturalne procesy degradacyjne zachodzące w XIX-wiecznych stalach zgrzewnych. Inżynieria i Budownictwo 69(11):610–612
24. Lesiuk G, Szata M (2010) Wpływ procesów degradacji mikrostruktur na udarność stali mostowych z przełomu XIX i XX w. Logistyka 6:1883–1892
25. Szata M, Lesiuk G, Pękalski G (2009) Assessment of degrading processes progress in the old brigde steel in terms of fracture mechanics-part one-the investigation of possibilities. Logistyka (6)
26. Raposo P, Duda M, Lesiuk G, Correia JAFO, Valente IB, Jesus AMP, Calçada RAB (2017) Structural degradation of puddle irons and steels from old riveted metallic bridges. In Atas do XI Congresso de Construção Metálica e Mista (pp 453–461). Associação Portuguesa de Construção Metálica e Mista (CMM)
27. Lesiuk G, Zięty A, Tabaszewski M, Rabiega J (2016) Diagnostyczna rola badań korozyjnych w detekcji procesów degradacji mikrostruktur zachodzących w długotrwale eksploatowanych stalach konstrukcyjnych. TTS Technika Transportu Szynowego, 23
28. Lesiuk G, Szata M (2011) Rozwój pęknięć zmęczeniowych w stalach zgrzewnych–opis procesów degradacji mikrostruktur. Zeszyty Naukowe. Mechanika/Politechnika Opolska 99:110–112
29. Lesiuk G, Pękalski G, Szata S, Fragassa C (2009) Monitorando la resistenza delle infrastrutture in acciaio. Fonderia e Pressofusione, pp 58–65
30. Gabetta G, Nykyforchyn HM, Lunarska E, Zonta PP, Tsyrulnyk OT, Nikiforov K, Vuherer T (2008) In-service degradation of gas trunk pipeline X52 steel. Mater Sci 44(1):104
31. Renowicz D, Hernas A, Cieśla M, Mutwil K (2006) Degradation of the cast steel parts working in power plant pipelines. J Achiev Mater Manuf Eng 18(1–2):219–222
32. Kryzhanivs'kyi EI, Nykyforchyn HM (2011) Specific features of hydrogen-induced corrosion degradation of steels of gas and oil pipelines and oil storage reservoirs. Mater Sci 47(2):127–136
33. Nykyforchyn H, Tsyrul'nyk O, Petryna D, Hredil M (2009) Degradation of steels used in gas main pipelines during their 40-year operation. Strength Mater 41(5)
34. Pękalski G (2004) Material aspects in the degradation theory of the surface mining machinery. In Mine Planning and Equipment Selection 2004: Proceedings of the thirteenth international symposium on mine planning and equipment selection, Wroclaw, Poland, 1–3 September 2004. CRC Press, p 479
35. Konat Ł, Pękalski G (2020) Overview of materials testing of brown-coal mining machines (Years 1985–2017). In Mining machines and earth-moving equipment. Springer, Cham, pp 21–58
36. Moroń JW (1979) Migracja węgla i azotu w niskich temperaturach w stalach ferrytycznych i martenzytycznych, Wydawnictwo Uniwersytetu Śląskiego, Katowice
37. Rasek J (2000) Wybrane zjawiska dyfuzyjne w metalach krystalicznych i amorficznych, Wydawnictwo Uniwersytetu Śląskiego, Katowice
38. Staub F, Adamczyk J, Cieślak Ł, Gubała J, Maciejny A, Metaloznawstwo (1994) Śląskie Wydawnictwo Techniczne, Katowice
39. Ferguson P (1985) Mechanical Properties of Nitrogen–Ferite. Metall Mater Trans 16(1), s 45–50
40. Skakov YA (1962) Structural changes during aging of commercial iron. Metal Csience Heat Treat 4(3–4):105–108
41. Muller J (1996) Untersuchungen zum Damungs und Alterungsverhalten vob Kohlenstoff und Sticstoff in einem weichen unberuhigten Thomasstahl, Aachen
42. Leda H (1998) Strukturalne aspekty własności mechanicznych wybranych materiałów, Wydawnictwo Politechniki Poznańskiej
43. Kossakowski P (2014) Load–carrying capacity of steel railway bridges subjected to long term service according to Eurocode standards. Roads Br Drogi I Mosty 13:115–130
44. PN-EN 1990:2004 Eurokod–Podstawy projektowania konstrukcji
45. PN-84/H-04334 (1984) Badania niskocyklowego zmęczenia metali
46. ASTM E606-04 (2004) Standard Practice for Strain-Controlled Fatigue Testing

47. Coffin LS (1954) A study of the effects of the thermal stresses on a ductile metal. Report, American Society of Mechanical Engineers
48. Manson SS (1954) Behaviour of aterials under conditions of thermal stress. Report NACA TN-2933, National Advisory Committee for Aeronautics
49. PN-74/H-04327 (1980) Badanie metali na zmęczenie, próba osiowego rozciągania-ściskania przy stałym cyklu obciążeń zewnętrznych
50. ASTM E466-07 Standard Practice for Conducting Force Controlled Constant Amplitude
51. ASTM E647-11 (2011) Standard test method for measurement of fatigue crack growth rates

Chapter 3
Sensitivity of Puddled Steels to Stress Corrosion Cracking and Estimation of Their State with Using Electrochemical Parameters

Abstract This chapter of book is devoted to the study of the old steels from the elements of different objects of the late XIX century in Poland such as the Sand Bridge, the Main Railway Station and the Central Pomorski Bridge. It was used the metallographic and fractographic studies, determination of mechanical characteristics during tensile and impact testing, stress corrosion cracking (SCC) using slow strain rate test (SSRT) and of the electrochemical investigations in the synthetic acid rain environment. It was shown the significant role of the microstructure of the puddled steels in their corrosion characteristics. Features of the failure mechanisms of puddled steels after tensile tests and impact tests and stress-corrosion cracking are described. Such studies had enable to uncover the mechanisms of destruction under controlled conditions of laboratory corrosion-mechanical tests and to establish the key role of the structural factor in ensuring the operability of old steels. It was shown the structural components from the puddled steels were especially inclined to SSC due to their very low levels of impact toughness. However, these steels continue to be exploited in the different structures. Therefore, it is very important to take into account their actual properties to justify the reliability of their further operation.

3.1 Introduction

Since the second half of the XIX century, puddles have become the most common material for metal structures, such as bridges across rivers, water towers, closed exhibition halls or railway stations. For example, among the most famous of them—the Eiffel Tower, many ancient bridges of Western Europe, as well as the original structure of the Statue of Liberty in the United States. Today the intensive repairs and renovation projects concerning to the old (XIX century) bridges are carried out in Poland. These bridges were made of puddled or cast steels with low carbon content. During long-term operation such steels are degraded on the microstructural level [1–3]. It is associated with the disappearance of pearlite areas, allocation of carbides inside the grains and numerous cementite particles along the grain boundaries. Such microstructure changes in turn significantly affect their mechanical properties.

G. Lesiuk et al., *Degradation Theory of Long Term Operated Materials and Structures*, Structural Integrity 15, https://doi.org/10.1007/978-3-030-43710-7_3

It is important for humanity to preserve such ancient constructions for future generations as an example of high achievements of human thought, the implementation of original engineering and technological solutions and interesting cultural and historical objects. However, the composition, structure and properties of the puddled steels can vary over a wide range [4–10]. It depended on their producers. Iron oxides (usually hematite) were used as a protective coating to prevent the combustion of molten metal in the furnace. As a result, the stability of the properties of puddled steels depended on subjective factors (related to the composition of ores used in melting) and objective ones (associated with the professional level of steel producers). Thus, even in the virgin state, puddles from different structural elements of the same structure can differ substantially from each other. It is also clear that long-term exploitation (over a hundred years) of ancient steels affected their technical state (through their degradation under the influence of power and climatic factors). The process of steel degradation is known to depend on both the stress level in structural elements [11] and climatic conditions (especially in technologically advanced regions) [12]. Both factors can significantly affect on the steels ability to resist workloads and thereby ensure the operability of ancient structures. Thus, in addition to the characteristics of steels in the virgin state and the degree of their degradation during long-term operation, the operational factors (such as the level of stresses and the environment composition), could play an important role in ensuring of their workability too.

3.2 Materials and Methods

The structural elements of the Sand Bridge (Wroclaw 1861), the Main Railway Station (Wroclaw 1855–1870) and the Central Pomorski Bridge (Wroclaw 1885) were used as the objects of study. All investigated structural elements were made of puddled steels. Three types of puddled steels from these ancient structures with significant difference in the content of the main elements were studied. Their chemical composition was determined on flat polished specimens with a working surface of 20×20 mm^2 and a thickness of 3...5 mm using the optical spark atomic emission spectrometer SPECTROMAX LMF 0.5. Average values from three measurements for each element in the steels are given in Table 3.1.

The prototypes of the exploited steels (for comparison) were chosen among modern carbon steels in the virgin state taking into account a compliance of their chemical composition (in terms of the content of the main elements and, first of all, the carbon content). The chemical composition of the selected prototype steels for the exploited puddled steels is also shown in Table 3.1. Therefore, in the hereinafter the characteristics of the exploited steels from three old objects will be compared with the properties of modern steels in the virgin state.

To assess the technical state of the exploited steels the metallographic and fractographic studies, strength and ductility characteristics at tension of specimens in air and in artificial acid rain at the stress corrosion cracking (SCC) using slow

Table 3.1 Chemical composition of the investigating steels after exploitation in elements of the old constructions (puddled steels) and in the virgin state (their modern prototypes)

Designation of steel	Type of steel	Objects	Content elements, wt%					
			C	Si	Mn	S	P	Cu
1	Puddled steels	Main Railway Station	0.07	0.52	0.15	0.028	>0.156	0.02
2		Sand Bridge	0.29	1.17	0.35	0.025	>0.156	0.01
3		Central Pomorski Bridge	0.40	1.17	0.45	0.111	>0.156	0.04
4	Modern steels	Armco Fe	0.09	0.08	0.34	0.024	0.013	0.10
5		St3	0.22	0.07	0.06	0.026	0.035	0.02
6		Steel 35	0.38	0.26	0.64	0.023	0.028	0.05

strain rate test (SSRT), impact toughness and electrochemical characteristics such as polarization resistance values were used.

The strength and plasticity characteristics of all steels were determined using UME-10T installation under uniaxial tension of specimens at their strain rate of the $3 \times 10^{-3}\ s^{-1}$ in air and of the $1.7 \times 10^{-7}\ s^{-1}$ in a model environment (artificial acid rain). Thus all analyzed steels were tested on the SSC using SSRT in the slightly acidic aqueous solution (artificial acid rain) with pH 4.4. The composition of this environment (in g/l: Na_2SO_4 (3.2); HNO_3 (1.58); $NaNO_3$ (2.12); NaCl (8.48), with the addition of H_2SO_4 to achieve pH 4.4 in the resulting solution) was chosen basing on the analysis of the rainwater composition in the industrial regions. The modern steel rolling in the form of cylindrical bars was used to make the cylindrical specimens of 5 mm in diameter from the steels in the virgin state. The flat specimens with a working cross-section of 4 × 4 mm made of puddled steels were cut out from real elements of old structures. Schematic drafts of both types of the specimens are shown in Fig. 3.1a, b. To exclude the influence of stress concentrators caused by machining of the working surfaces of specimens and thereby reduce the scatter of the obtained data, the surfaces of all tensile specimens were polished before testing using pastes of different granularity.

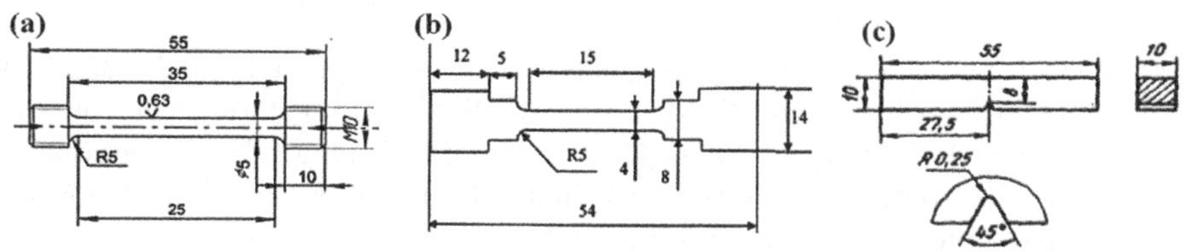

Fig. 3.1 The specimens used **a**, **b** under uniaxial tension in air and during stress corrosion cracking (SCC at SSRT) tests in artificial acid rain and **c** in the Charpy impact test

The impact tests of Charpy specimens were carried out using the IO-5003 installation. Schematic draft of used specimens is shown in Fig. 3.1c. The cross-sections of all tested specimens were 10 × 10 mm^2. Only the thickness of specimens from the Sand Bridge metal did not meet the requirement to the specimen size, it was only 7 mm (instead of the required 10 mm).

The corrosion resistance of analyzed steels was estimated using the results of electrochemical tests obtained by the potentiodynamic method. As an electrolyte the aqueous solution (its composition is described above as the artificial acid rain) was used at temperature of 298 K. All electrochemical tests were carried out on potentiostat IPC-Pro. A standard electrochemical cell with a typical three-electrode system (the analyzed steel as a working electrode, a reference electrode (saturated Ag|AgCl) and an auxiliary Pt electrode) was used. The plates with dimensions of 15 × 15 mm and thickness ~3 mm were cut from analyzed steels and used as the working electrodes. Before electrochemical studies all surfaces of working electrodes were polished and covered with an insulating waterproof coating. Only the area of ~0.4 cm^2 had not been covered by insulating coating.

The method of polarization resistance R_p was realized by polarization of the working electrode in the vicinity of the corrosion potential (within the range ±30 mV in the anodic and cathodic directions respectively). For many systems metal—environment the linear dependency of the polarization potential E versus polarization current density j exists in this range of potentials (so sometimes this method is also called the linear polarization method). However, more often the linear part of the polarization curve is completely absent or exists in a very narrow range of potential values. In this case, a tangent line to the polarization curve should be drawn at a point corresponding to the corrosion potential E_{cor} (at the corrosion current density $j_{cor} = 0$). The slope of the tangent line determines the value of the polarization resistance $R_p = \Delta E / \Delta j$.

The Neofot-21 optical microscope and the EVO-40XVP scanning electron microscope with a micro-X-ray spectral analysis system INCA Energy 350 were used in the metallographic and fractographic studies.

3.3 The Results and Discussion

3.3.1 Analysis of the Composition of Old Steels

As follows from Table 3.1, the carbon content in analyzed puddled steels varies, in wt%, from 0.07 (in steel from the Main Railway Station) to 0.29 (in steel from the Sand Bridge) and 0.4 (in steel from the Central Pomorski Bridge). The silicon content in these steels also differs more than two times (from 0.52 in steel from the Main Railway Station to 1.17 in steels of other analyzed objects). The manganese content also varies from two to three times (from 0.15 in steel from the element of the Main Railway Station to 0.35 and up to 0.45 wt% in steel from the Sand Bridge and in steel from the Central Pomorski Bridge, respectively). An extremely high content of sulphur and phosphorus in steel from the Central Pomorski Bridge (0.11 and more than 0.16 wt%, respectively) was also noted, as compared with 0.03 and 0.02 wt% in steels from other analyzed old objects. So, notable differences in the content of the main components in puddled steels are assumed to be a result of differences in ore composition used during melting, and the differences in techniques used, etc. It is evident too that the compositions of analyzed puddled steels and their modern prototypes by the content of main components (especially in terms of carbon content) are in satisfactory agreement (Table 3.1). It was conditionally accepted that in mostly the Armco iron (steel 4 according to Table 3.1) can be the prototype for the steel from the Main Railway Station (steel 1), and the St3 steel (steel 5)—for the steel from the Sand Bridge (steel 2), and the steel 35 (steel 6)—for the steel from Central Pomorski Bridge (steel 3).

Undoubtedly, differences in the composition of puddled steels will effect on their mechanical properties (especially under the influence of a corrosive environment). It is well known that the strength level of steel increases, but the impact toughness and plasticity value decrease with rising of carbon content. Technological impurities such as manganese, silicon, sulphur and phosphorus are inevitable components of steels. It is difficult to remove them during both melting (P, S), and deoxidation (Si, Mn). Silicon is completely soluble in ferrite at a content of up to 0.30 wt% and increases the strength of ferrite, but almost does not reduce its plasticity. Moreover, the smaller the ferrite grains, the higher the strength of steel. But with an increase in silicon content in general-purpose carbon steels more than 0.40% their ductility can be significantly reduced. However, namely such the higher silicon content has been found in all three analyzed puddled steels. Manganese neutralizes the harmful effect of sulphur by formation of sulphides. In addition, Mn can dissolve in cementite, partially replacing iron atoms in its crystal lattice. In this case, doped cementite $(Fe, Mn)_3C$ forms, which practically does not differ in properties from Fe_3C.

The range of change in sulphur content in the steels (as the technological impurity) usually are 0.035–0.06 wt%, and in phosphorus 0.025–0.045 wt%. These elements are the most harmful impurities in steels. Even a slight increase in their content in steels reduces their plasticity, impact toughness and corrosion resistance. The harmful effect of phosphorus on the steel's properties is stronger at the higher carbon content in it. The negative effect of phosphorus on the properties of steels is associated with its unlimited solubility in liquid iron, but low solubility in solid one. During crystallization and cooling of steel, an excess amount of phosphorus in the form of phosphides release from a supersaturated solution. Their melting point is lower than the crystallization temperature of a metal. As a result melted phosphides mainly locating along the grain boundaries reduce the ductility and impact toughness of the steels (especially at low temperature). Thus, if the content of sulfur in the analyzed steels is within the permissible limits, then the significantly exceeded content of phosphorus in them is very dangerous.

3.3.2 Microstructure of Old Structures Steels

A large number of nonmetallic inclusions oriented along the rolling direction were observed even at low magnification of the light microscope on the non-etched metallographic specimens of all puddled steels (in longitudinal and transverse sections with respect to the rolling direction of structural elements). These inclusions are more difficult to polish and therefore they look darker in reflected light than the metal around them (Fig. 3.2). The discontinuous chains of nonmetallic inclusions along the rolling direction were clearly expressed in the longitudinal sections of all steels (Fig. 3.2a–c), but in their transverse sections this feature was less obvious (Fig. 3.2d). As it will be shown further by the micro-X-ray spectroscopy, these nonmetallic inclusions were mainly iron oxides (due the interaction of ore residue with carbon had not finished during steels deoxidation).

Significant irregularity of slag stringers distribution over the thickness of the elements is also noted as a feature of all puddled steels. Their shape, length and cross-sectional area varied widely. As a result their contact area with steel matrix varied significantly and it could influence the mechanical properties of steels. It is clear that cohesion along these interphase boundaries is very important for ensuring the

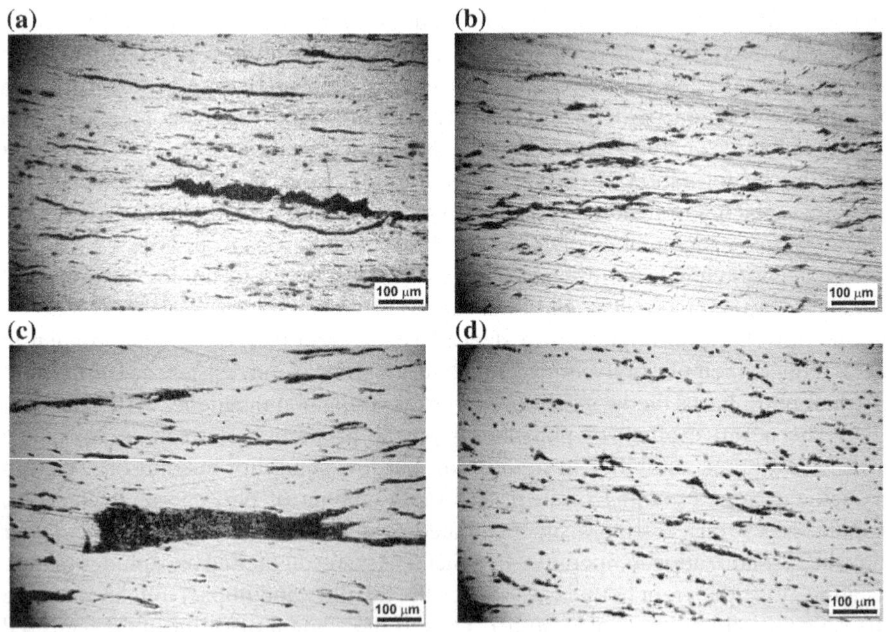

Fig. 3.2 Metallographic view **a**–**c** of longitudinal and **d** transverse with respect to rolling direction non-etched sections of the puddled steels cut from the structural elements **a** of the Main Railway Station, **b** the Sand Bridge and **c**, **d** the Central Pomorski Bridge

required level of strength and plasticity for these steels and their ability to counteract the long-term impact of corrosive environments.

The microstructures of the analyzed puddled steels significantly differed after etching of their polished sections with Nital. The ferrite grains with polygonal morphology prevailed in the structure of the steel (1) from the Main Railway Station (Fig. 3.3a). A typical ferrite-pearlite structure was revealed on the steel (2) from the elements of the Sand Bridge (Fig. 3.3b, c). The ratio of ferrite and pearlite in the structure of this steel varied significantly along the structural elements thickness. The maximum amount of pearlite in the steel was found in the central part of the cross section of a bridge element (Fig. 3.3b), and the minimum one—near its both surfaces (Fig. 3.3c). The pearlite content in these zones could differ up to three times.

The most unusual structure was revealed on steel (3) from the Central Pomorski Bridge. Taking into account the highest carbon content in this steel, pronounced ferrite-pearlite structure was expected here, similar to that revealed on the steel from the Sand Bridge. However, these expectations were not justified. Ferrite grains predominated in the structure of the steel (3), and a small amount of pearlite grains was located along the grain boundaries of the ferrite in the form of narrow interlayers (Fig. 3.3d). However, in contrast to the steels already analyzed, a large number of round or slightly elongated inclusions were observed in the steel (3) from this bridge element. It was assumed that a significant part of these inclusions are carbides,

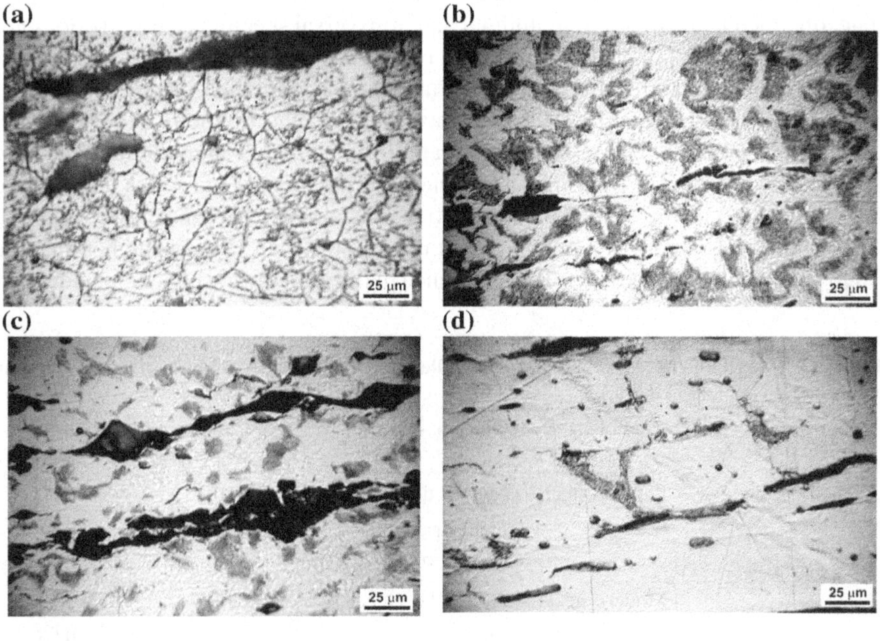

Fig. 3.3 The microstructures on the etched metallographic sections (longitudinal relative to the rolling direction) of the puddled steels cut from the structural elements **a** of the Main Railway Station, **b**, **c** the Sand Bridge, **d** the Central Pomorski Bridge

which provide such high content of carbon in this steel. This assumption will be later confirmed by the results of the micro-X-ray spectroscopy. The fine-dispersed inclusions of this form were mainly iron carbides alloyed by the manganese. Their presence in the steel structure may be due to the incomplete interaction of ledeburite with iron ore in the process of smelting of the puddled steel. As a result, these carbides remained visible in the structure of steel (3) against the background of iron matrix.

3.3.3 Impact Toughness of Puddled Steels

It is known that general-purpose steels in the virgin state with higher carbon content is characterized by lower impact toughness. The results of present studies of modern steels not been in operation confirmed this conclusion (Table 3.2, steels 4–6). Analogically, similar effect of carbon content in the puddled steels on their resistance to brittle fracture was expected after their long-term operation. However such clear dependency between carbon content and impact toughness has not been obtained for previously operated puddled steels (Table 3.2, steels 1–3).

The minimum value of impact toughness was obtained for the steel (1) with minimum carbon content. It was assumed this could be concerned with the transverse orientation of the specimens relative to the rolling direction of structural elements. However, the difference between impact toughness values of the metal in the virgin state, obtained by testing of longitudinal and transversal specimens, usually does not exceed 12…0.50% [14–16]. The steel (1) can be compared (in the first approximation) with a technical iron (Armco iron) if take into account their likeness for carbon content and ferrite structure. However, according to the literature [13], Armko's value of the impact toughness is more than 30 times higher than that obtained for the exploited steel (1). The impact toughness of the steel (4) taken as a prototype for the steel (1) of the Main Railway Station was almost two times lower than the value obtained for modern Armco iron but it was more than 15 times higher than the value obtained for the steel (1). The difference in the modern and old compared steels

Table 3.2 Impact toughness of the analyzed puddled steels and their modern prototypes, depending on the carbon content in them

Designation of steel	Type of steel	Objects, steels	Carbon content, wt%	KCV, MJ/m^2
1	Puddled steels	Main Railway Station	0.07	0.06
2		Sand Bridge	0.29	0.45
3		Central Pomorski Bridge	0.4	0.25
4	Modern steels	Armco-Fe	0.09	1.13
5		St. 3	0.22	0.98
6		Steel 35	0.38	0.46
[13]		Armco-Fe*	0.025	2.0

consists in a large number of nonmetallic inclusions presented in the structure of the exploited steel from the Main Railway Station (Fig. 3.2a). As it will be shown further by micro-X ray spectral analysis these slag stringers are iron oxides. It was assumed that such a low impact toughness value for puddled steel from the element of the Main Railway Station was associated with a large number of these slag inclusions. In addition, it is possible that brittle fracture resistance of ferrite (puddled steel has practically ferrite structure) could degrade during long-time operation.

The resistance to brittle fracture of the puddled steel (2) cut from the Sand Bridge element is almost two times lower than of the modern prototype of this steel St3 (5, Table 3.2). A similar two-fold decrease in the impact toughness was recorded for the steel (3) as compared to its modern prototype (6). Ignoring of the slight discrepancy between the old and modern steels in composition and structure (in particular, the heterogeneous distribution of perlite in the thickness of the steel element (2) and the longitudinal chains of large slag inclusions in the structure of old steels) such a significant decrease in the toughness of puddled steels should be primarily associated with their operational degradation.

3.3.4 Fractography of the Charpy Specimens

Typical signs of transgranular brittle fracture were revealed on the specimens fracture surfaces of the steels (1) and (4) after their impact tests (Fig. 3.4). Only several small areas (up to 100 μm^2) with ductile dimples relief were observed just near the stress concentrator but only on a sample of the steel (4) (Fig. 3.4b).

It is consistent with a higher impact toughness of this steel. Moreover, ductile fragments on the fracture surfaces were found in both steels but only at the large-angle boundaries between adjacent grains in their structure (Fig. 3.4a, b, d). Such ductile elements were larger at the fracture surface of the steel (4). The fracture features of the steel (1) were the following. Firstly, a large number of slag inclusions along the direction of fracture propagation were presented on the surface (Fig. 3.4a, c). Secondly, their significant decohesion from the matrix and destruction of the bridges between the nearest inclusions due to shear deformation was observed (Fig. 3.4c). Thirdly, the transgranular fracture along conglomerates of grains due to a slight disorientation of atomic rows in adjacent grains was noted (Fig. 3.4a, c). Whereas the direction of failure on the fracture surface of the steel (4) varied continuously from a grain to a grain. The evidence of that was a clear change in the orientation of the river patterns (as signs of cleavage fracture) during crossing grain boundaries (Fig. 3.4b, d).

Nonmetallic inclusions observed on the background of the flat cleavage facets of the fracture surface of the steel (1) were subjected to the micro-X-ray spectral analysis. It was shown that all inclusions were the slag remains (Fig. 3.5). All the elements found in the steel (1) were presented in the composition of nonmetallic inclusions (regardless of their shape). Oxygen content in small and round inclusions (up to 1 μm in diameter) was higher, and it was less in larger ones elongated in the

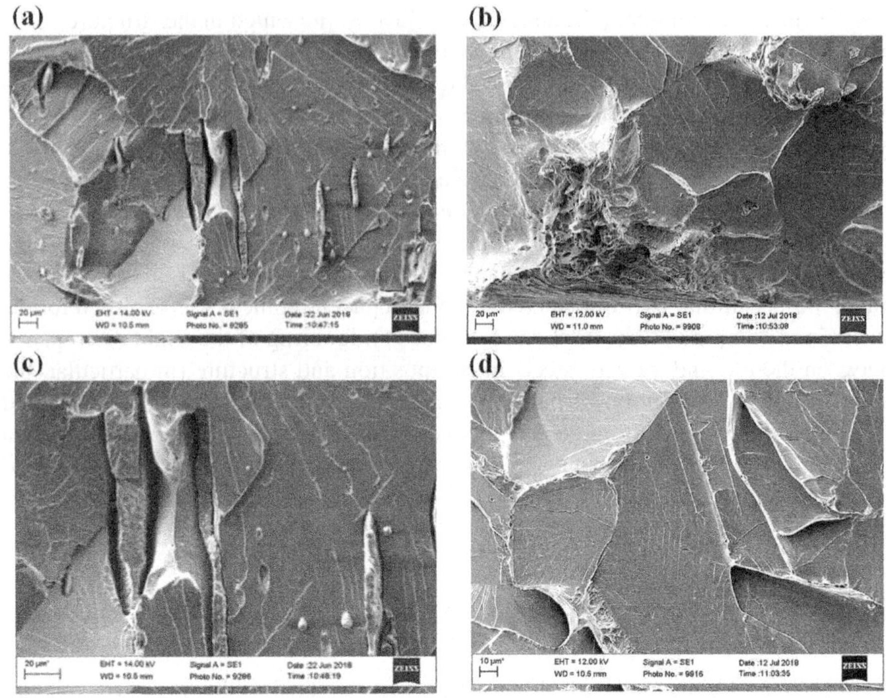

Fig. 3.4 Fractographies obtained on the fracture surfaces of impact specimens **a**, **c** of the steel (1) cut from element of the Main Railway Station and **b**, **d** steel (4) used as it prototype

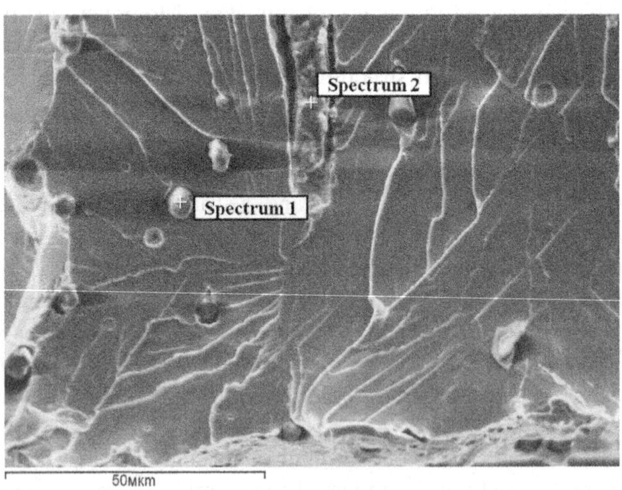

Spectrum 1

Element	Wt %	At %
O K	31.10	56.46
Si K	7.47	7.72
P K	9.20	8.63
Mn K	1.81	0.96
Fe K	50.42	26.23
Totals	100.00	

Spectrum 2

Element	Wt %	At %
O K	7.12	19.82
Si K	2.91	4.61
P K	1.50	2.16
S K	4.70	6.54
Mn K	2.60	2.11
Fe K	81.17	64.76
Totals	100.00	

Fig. 3.5 The micro-X-ray spectral analysis of the nonmetallic inclusions revealed on the fracture surface of the steel (1) after the impact test

fracture direction. But the iron content was significantly higher in large inclusions. Since the destruction of a specimens are usually occurred along the most weakened places in its cross section, so low impact toughness of the puddled steel (1) can be associated with inclusions as slag residues. Their bond with the matrix has been weakened by damages appeared during the operation of the steel. These inclusions with weak cohesion with the matrix divided the working section of specimen into thin fragments in the form of bridges from undamaged metal between these inclusions. Externally (on a macro scale), the steel looks like a solid metal without visible damages, but on the micro scale level in the cross section of specimen the network of damages oriented in the failure direction were formed. These damages, by decreasing the actual working cross-section of the samples, reduce their ability to resist of the external loads. Thus, the resistance to brittle fracture of steel (1) has significantly decreased due to the developed network of structurally caused damages.

The impact toughness of the steel (2) from the element of the Sand Bridge was the highest among the analyzed versions of operated puddled steels. Ductile fracture mechanism dominated on the fracture surface of the impact specimen (Fig. 3.6a, b). Many delaminations initiated by nonmetallic inclusions were also observed in this puddled steel (Fig. 3.6a). These delaminations divided the specimen fracture surface into bands of different width, oriented in the failure direction. The bridges between the nearest delaminations were destroyed in a ductile manner. It was noted that big dimples initiated at the interface of small pearlite grains with the ferrite ones. As a result the pearlite was clearly visible on the bottom of most dimples (Fig. 3.6c). The bridges between adjacent pores formed at the interface of two phases were destroyed by tension up to destruction. Their stretching within the ferrite grains was observed.

Cleavage fracture with typical river patterns was observed only in a narrow band oriented along the direction of fracture propagation (Fig. 3.6e). The cleavage facets propagated into the depth of the fracture surface to a distance 2…0.5 times bigger than a grain size of ferrite. Consequently, cleavage fracture propagated by united front through the conglomerate of grains with a close arrangement of atomic planes favorably disposed for cleavage. Ferrite grains predominated in this band while pearlite ones practically were not identified. Hence, it was concluded that fracture of the operated steel (2) by realization of a typical cleavage mechanism is possible only within zones with small pearlite content.

Fracture mechanism of the specimen from the steel (4), accepted as a modern prototype of the puddled steel (2) was predictable. Ductile fracture in the vicinity of the stress concentrator has changed to cleavage (Fig. 3.6b, d, f). However, in this case flat dimples of the parabolic shape were formed by shear and dominated in the zone of ductile fracture (Fig. 3.6b), whereas in the case of the steel (2) such dimples were formed by stretching of ferrite bridges between pores (Fig. 3.6a). Features of the brittle part of this fracture surface were noted, namely, local directions of failure varied from grain to grain, and the river patterns due to cleavage were revealed in the body of ferrite grains (Fig. 3.6f).

The micro-X-ray spectral analysis of the nonmetallic inclusions observed on the background of the fracture surface of the steel (2) confirmed that all inclusions were the slag remains (Fig. 3.7). All the elements found in the steel (2) were revealed in

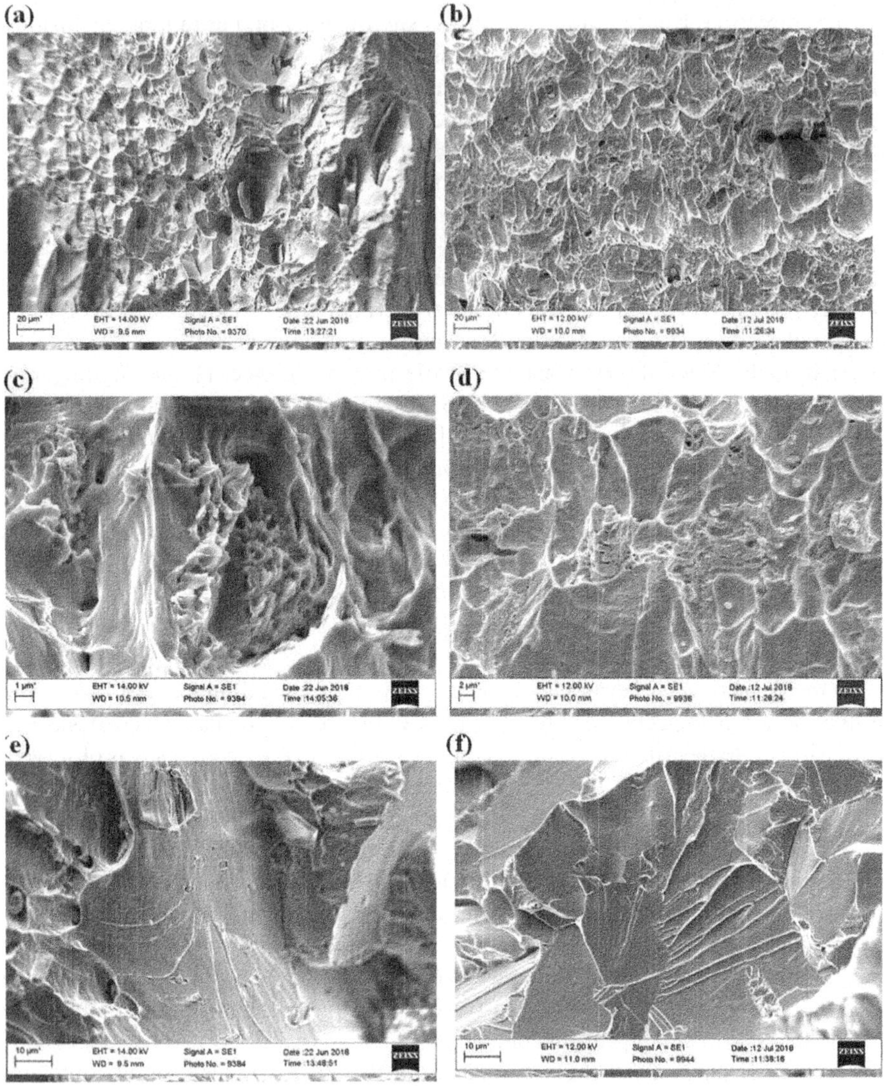

Fig. 3.6 Fracture surfaces of impact specimens **a**, **c**, **e** of the steel (2) cut from element of the Sand Bridge and **b**, **d**, **f** the modern steel (5) used as it prototype

the composition of nonmetallic inclusions (regardless of their shape). Thus, it was confirmed that the main feature of both puddled steels is the presence of the remains of slag in it that had not finished reacting with the melt.

Fractographic features on the fracture surface of the steel (3) are mainly associated with structure peculiarities. Large and practically round dimples formed around the primary carbides were observed on the sections with ledeburite structure (Fig. 3.8a,

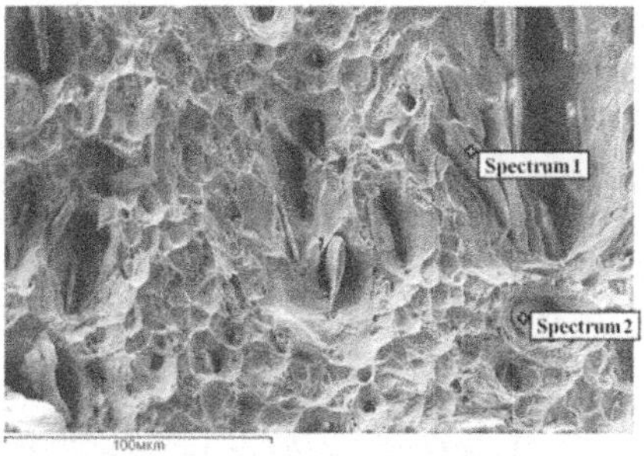

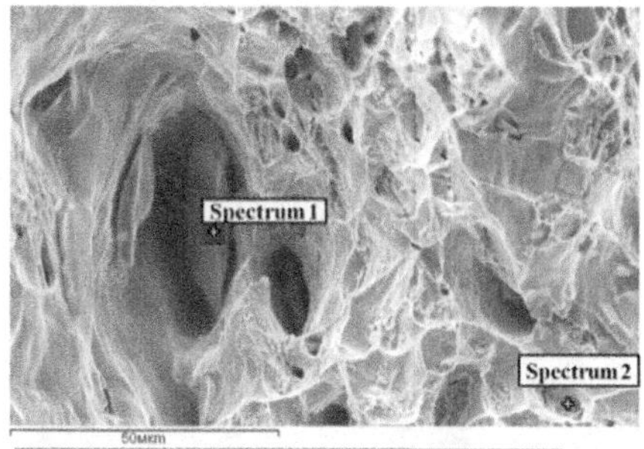

Spectrum 1:	Element	Wt %	At %
	C K	3.63	14.76
	Si K	0.84	1.46
	Mn K	10.77	9.58
	Fe K	84.77	74.20
	Total	100.00	

Spectrum 1:	Element	Wt %	At %
	O K	17.06	41.07
	Si K	2.53	3.47
	Fe K	80.41	55.46
	Total	100.00	

Spectrum 2:	Element	Wt %	At %
	C K	19.97	50.36
	O K	4.47	8.10
	Si K	1.41	1.30
	Fe K	74.17	40.23
	Total	100.00	

Spectrum 2:	Element	Wt %	At %
	O K	15.53	38.25
	Si K	3.05	4.29
	Fe K	81.42	57.46
	Total	100.00	

Fig. 3.7 The micro-X-ray spectral analysis results of the nonmetallic inclusions revealed on the fracture surface of the steel (2) after impact test

c). In the regions with ferrite structure, the destruction occurred by means of transgranular cleavage (Fig. 3.8c). Many large nonmetallic inclusions oriented along the direction of failure were observed on the fracture surface of the steel (3) as well as in the both previous versions of puddled steels (1 and 2). Long delaminations were initiated from these inclusions (Fig. 3.8a, c).

At the same time, the ductile fracture mechanism prevailed at the fracture surface of the modern steel (6) in the vicinity of the stress concentrator. The pearlite grains were observed at the bottom of many parabolic dimples formed by shear (Fig. 3.8b). However at a distance of ~420 μm from the bottom of the concentrator the relief by the parabolic dimples was replaced by the facets of the transgranular cleavage (Fig. 3.8d). Within this part of fracture surface pearlite grains were visualized in the form of narrow ductile zones along the boundaries of the transgranular cleavages within ferrite grains.

The micro-X-ray spectral analysis of nonmetallic inclusions observed against the background of the fracture surface of steel (3) confirmed that most of them were large particles of the remaining slags with the irregular shapes (Fig. 3.9). However, a large number of fine particles have been identified as compounds of the iron-carbon type (primary iron carbides with manganese additions) that have not been removed from

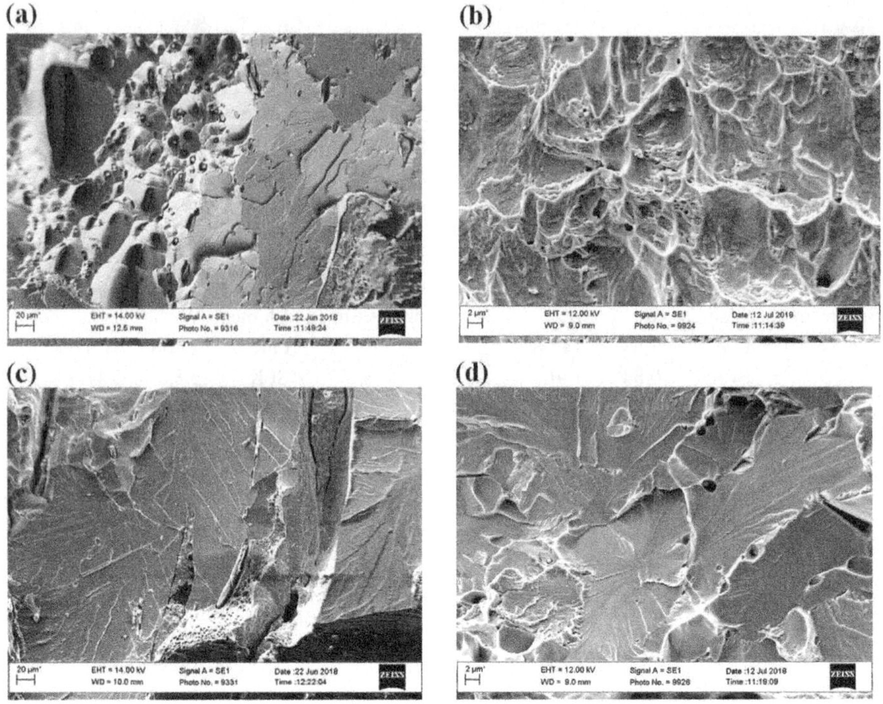

Fig. 3.8 Fracture surfaces of impact specimens **a**, **c** of the steel (3) cut from the element of the Central Pomorski Bridge and **b**, **d** the modern steel (6) used as its prototype

the puddled steel during its manufacture. Thus, the presence of largest accumulations of slag residues on it fracture surface that had not reacted with the melt is the main feature of this version of puddled steel too. Moreover the maximum carbon content in steel (3) was allocated in the form of the carbides particles in a ferrite matrix.

3.3.5 Tensile Test of Smooth Specimens in Air

Mechanical characteristics of the analyzed steels were estimated under tensile tests of the smooth specimens in air and in artificial acid rain. Typical stress-strain curves for the specimens from different steels tested in air and in corrosive environment are shown in Fig. 3.10. Strength (ultimate strength σ_{UTS} and yield strength σ_{YS}) and plasticity (elongation and reduction of area) characteristics obtained in air are shown in Table 3.3. The strength characteristics of the modern steels for general use usually increase while the plasticity characteristics decrease with rising of carbon content in them. This dependence is also confirmed by our tests of modern steels (5) and (6).

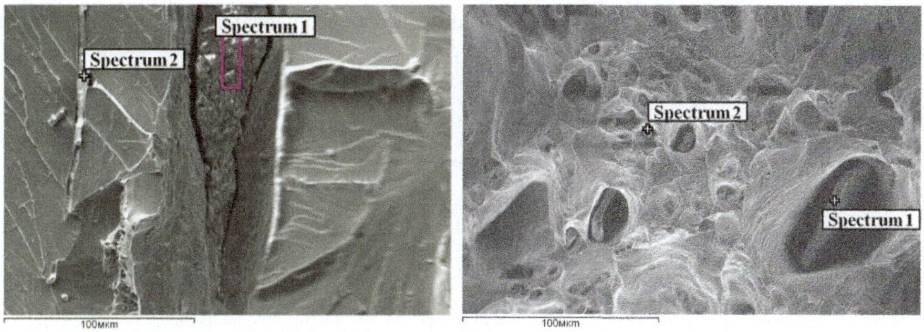

Spectrum 1:	Element	Wt %	At %
	C K	2.51	10.65
	Si K	0.12	0.22
	P K	0.33	0.54
	S K	0.10	0.16
	Fe K	96.94	88.42
	Total	100.00	
Spectrum 2:	Element	Wt %	At %
	O K	27.79	54.16
	Si K	7.41	8.23
	P K	3.12	3.14
	Mn K	3.24	1.84
	Fe K	58.44	32.63
	Total	100.00	

Spectrum 1:	Element	Wt %	At %
	C K	5.25	20.33
	Si K	0.39	0.65
	P K	0.37	0.56
	S K	0.26	0.38
	Fe K	93.73	78.08
	Total	100.00	
Spectrum 2:	Element	Wt %	At %
	C K	5.72	21.89
	Si K	0.15	0.25
	P K	0.42	0.63
	S K	0.05	0.08
	Fe K	93.66	77.15
	Total	100.00	

Fig. 3.9 The micro-X-ray spectral analysis results of the nonmetallic inclusions revealed on the fracture surface of the steel (3) obtained after impact test of specimen

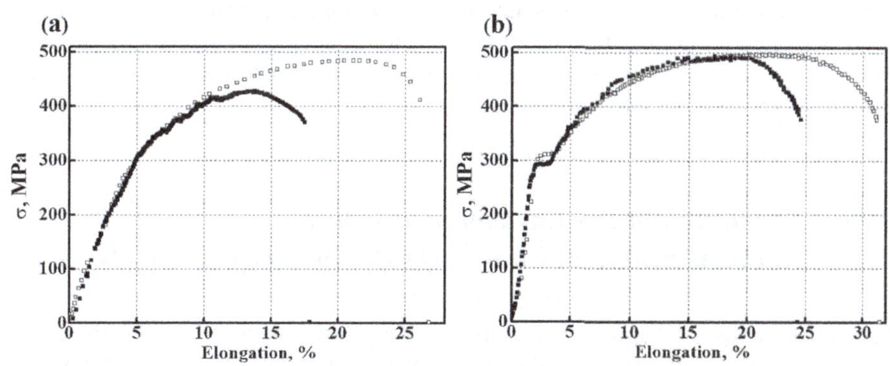

Fig. 3.10 Typical stress-strain curves for the specimens **a** of the steel (2) from the Sand Bridge element and **b** of the modern steel (5) tested in air (*white dots*) and in corrosive environment (*black dots*)

Table 3.3 Mechanical characteristics of analyzed steels obtained under tensile tests of the smooth specimens in air

Designation of steel	Type of steel	Objects, steels	Carbon content, wt%	σ_{UTS}, MPa	σ_{YS}, MPa	σ_{YS}/σ_{UTS}	Elongation, %	RA, %
1	Puddled steels	Main Railway Station	0.073	335	205	0.63	11.3	23.0
2		Sand Bridge	0.288	483	275	0.57	21.0	33.0
3		Central Pomorski bridge	0.399	414	241	0.58	27.9	31.8
5	Modern steels	St.3	0.224	497	310	0.62	28.3	61.5
6		Steel 35	0.382	653	390	0.59	22.4	54.5

However, this relationship between the mechanical characteristics and the carbon content in puddled steels was not observed. For instance, in the puddled steel (3) with the maximum carbon content, the strength characteristics had average values (compared to the values for steels (1) and (2), respectively). At the same time plasticity characteristics of this steel were the highest (Table 3.3). In addition, it is difficult to believe that the steel (1) with minimum carbon content and predictably the lowest strength (based on commonly accepted relationships between these indicators) has minimal plasticity. Thus, the relationship between carbon content and mechanical characteristics, well known for modern steels, has not been confirmed in the case of puddled steels tested in air. The reason of that could be two factors. The former may be related to the differences in the structures of these steels (number, size and shape of nonmetallic inclusions, the presence and morphology of pearlite grains, etc.). The latter is related to the peculiarities of degradation processes of steels with different carbon content. To determine the role of each one, further testing of all steels was carried out in acid rain environment.

3.3.6 Estimation of the Environment Effects on the Mechanical Characteristics of Analyzed Steels

The change in the mechanical characteristics of analyzed steels obtained in the corrosive environment with respect to the corresponding characteristics in air is shown in Fig. 3.11. The values λ characterize deviations of analyzed characteristics (strength and plasticity) under the influence of the environment (acid rain) comparing with the corresponding characteristics obtained in air. It was concluded that both strength and plasticity characteristics of all tested steels are reduced by the action of acid rain. The effect of corrosive environment on the characteristics of modern steels is

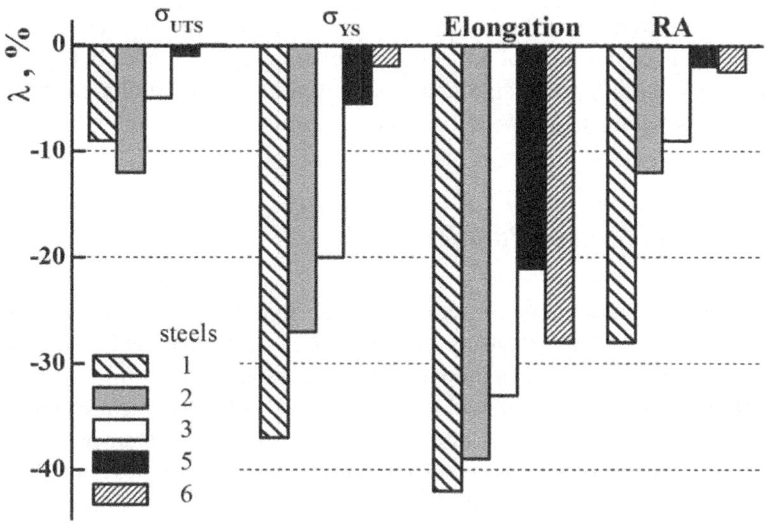

Fig. 3.11 The influence of the environment (artificial acid rains) on the mechanical characteristics of the investigated steels after SSC with SSRT, expressed by the value of ratio λ determined as the deviations of the correspondent characteristics under the influence of the environment to the obtained under the tests in air (steel designation, as in Table 3.3)

significantly less than on the corresponding characteristics of puddled steels. The environment effect on strength of modern steels is negligible (less than 10%). Its maximum effect was found in the modern steel (6) with a 28% reduction in elongation. Thus, under higher carbon content in the modern steels (and more pearlite in their structure, respectively), their elongation significantly reduced under the influence of the corrosive environment. Pearlite content could play a key role in this effect.

It was noted that the influence of the corrosive environment on both plasticity characteristics of puddled steels is not consistent with carbon content in them (namely, the higher carbon content, the weaker the environment effect on these characteristics). The effect of carbon content on plasticity characteristics of puddled steels was opposite to that observed in the modern ones. Therefore, the obtained inversion of carbon content effects on plasticity of the exploited puddled steels and non-exploited modern steels could be considered as a special manifestation of steel degradation during long-term operation. It is also noted that such single-valued inverse relationship between plasticity characteristics and carbon content was not found under the testing of puddled steels in air. No effect of carbon content (direct or reverse) on impact toughness of the puddled steels had also been observed (Table 3.2). Consequently, the relationship between the degradation of puddled steels and their carbon content was manifested only at their testing in the corrosive environment.

As a feature of puddled steels, unexpectedly strong effect of the corrosive environment on their yield strength was noted (Fig. 3.11). Such a strong decrease in

σ_{YS} under the influence of the environment (in particular for the steel (1) to 38%) essentially reduced the stage of uniform deformation of the samples before its localization. As a consequence, a significant reduction in both plasticity characteristics was observed in comparison with that obtained for the modern steels.

3.3.7 Degradation of Puddled Steels

The degradation effect was represented by the values of ratio λ characterized the strength deviation and the ductility characteristics of the puddled steels from the corresponding characteristics of modern steel (5) obtained after testing in air and after SSC with SSRT in an artificial acid rain (Fig. 3.12). All mechanical characteristics of both puddled steels were invariably lower compared to the characteristics of the modern steel. Thus, this is a common sign of the degradation of these puddled steels.

Besides, the degradation of both analysed puddled steels is more pronounced when it expresses using characteristics obtained in the corrosive environment (during SSC with SSRT), compared with the corresponding characteristics obtained in air. In addition, reduction in area is the most sensitive to the degradation of these steels. It should be noted that the estimation of the degradation level of heat-resistant steels (after long-term operation on steam pipelines of the HPP at high temperature and stress) has been carried out earlier [17–19]. It was shown that elongation was more sensitive to degradation of these steels. This was associated with opening of operational damages within the specimen cross-section. As a result, elongation of specimens from the damaged metal increased additionally (not only because of metal deformation in the working cross-section of a specimen, but also because of opening of these damages) [20]. This aspect was the feature of the degradation of heat-resistant steels. In the case of puddled steels, it can be assumed that, in addition to

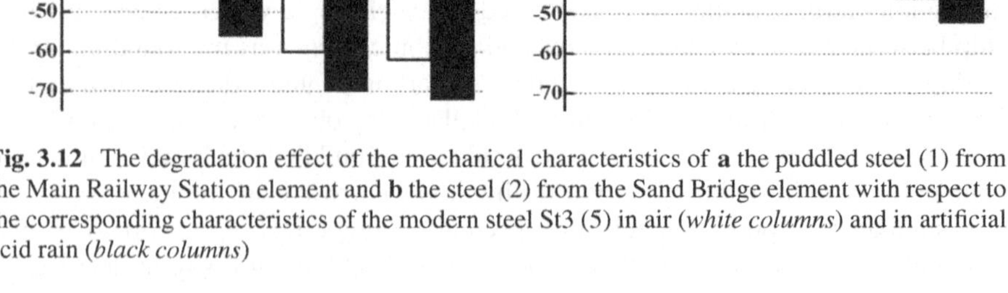

Fig. 3.12 The degradation effect of the mechanical characteristics of **a** the puddled steel (1) from the Main Railway Station element and **b** the steel (2) from the Sand Bridge element with respect to the corresponding characteristics of the modern steel St3 (5) in air (*white columns*) and in artificial acid rain (*black columns*)

above mentioned manifestation of degradation due to opening of damages in operated steels (in the form of long interfacial delamination), accelerated localization of failure process became possible in the most damaged section of specimens. These phenomena manifested themself even more under the stress corrosion cracking test of degraded puddled steels. It was assumed that the corrosive environment can easily seep through the metal, using the delamination along interphase boundaries as the channels that already existed in the cross section of specimens before their testing and opened during their deformation. If this assumption is correct, then the corrosive environments can practically unrestricted seep through the metal. Taking into account favourable conditions for electrochemical corrosion due to differences in the electrode potentials of the microstructural components of puddled steels, the corrosion damages will be propagated to increasingly deeper layers of cross-section of elements of old structures from these steels. As a result, under seeming external integrity of these elements their cross sections are transformed into a porous metal (pumice-like). Their resistance to combined action of external loading and aggressive rainfall in industrial regions should be grounded by additional studies (in particular, the static and cyclic crack growth resistance of exploited steels in a corrosive environment) and calculation of the strength and integrity of structural elements and structures as a whole with taking to account of obtained results.

3.3.8 Fractography of the Specimens Tested in Air and Artificial Acid Rain

A common feature of the destruction of all analyzed steels (puddled and their modern prototypes) was the ductile fracture of specimens subjected to tension in air (Fig. 3.13). Such fracture mechanism realizated by pores formation on the nonmetallic inclusions, carbides or along interfaces between ferrite and pearlite grains. The bridges between adjacent pores are stretched (until their break) during tension of specimens. In all cases the bridges between pores were located in the ferrite matrix. As a result dimples formed on the fracture surfaces due to coalescence of neighboring pores. Dimensions and shapes of these dimples depended on the morphology of inclusions or pearlite grains. In the central part of the fracture surfaces of the modern steels the bridges between dimples were usually broken due to the action of the normal stress component with formation of equiaxial dimples (Fig. 3.13d). However in puddled steels signs of the predominance of the shear stress component in their rupture were observed (Fig. 3.13b). This is evidenced by the parabolic shape of dimples dominated not only in the vicinity of the side surfaces of specimens (as it is usually seen under the ductile fracture of smooth specimens), but also in the central part of their fracture surfaces.

It was assumed that the degradation of puddled steels had already occurred during the operation of puddled steels in the elements of old constructures due to decohesion (at least partial) of nonmetallic inclusions from the ferrite matrix. As a consequence,

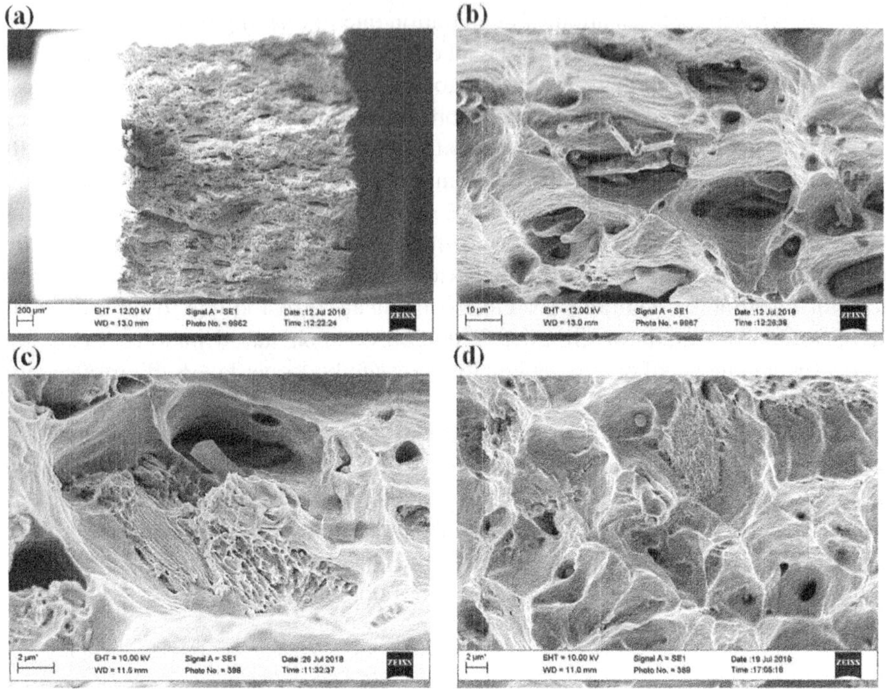

Fig. 3.13 Fracture surfaces **a, b** of the steel (3) from the Central Pomorski Bridge element, **c** of the steel (2) from Sand Bridge element and **d** the modern steel (5) corresponding to the steel St3, obtained under SSRT in air

deep holes around the inclusions formed on the fracture surfaces. Thus free surfaces arisen along interfaces between inclusions and the matrix (Fig. 3.13b, c). It indicates a weak cohesion of inclusions with the matrix in the exploited puddled steels, and it can be one of fractographic features of the degradation of these steels. Dislocation slip bands could easily exit on the already existing free surfaces along the interfaces between inclusions and matrix in puddled steels during their deformation (Fig. 3.13b). In the same way, dislocation slip bands exit to the outer surface of the specimens under loading. Thus practically parallel traces due to the exit of slip bands on the free surfaces around inclusions can indicate their existence already at the start of the specimens loading.

Significant texturing of the fracture surfaces of puddled steels was also noted at the testing in air (Fig. 3.13a). It concerned with the arrangement of nonmetallic inclusions along the deformation axis of structural elements during their production. Substantial part of these inclusions remained at the bottom of dimples formed around them (Fig. 3.13b). Thus, the bond between inclusions and ferrite matrix was retained (at least partially). Even small cementite particles inside the pearlite grains retained a contact with the matrix (Fig. 3.13c).

Fractographic features of SSC of puddled steels were analyzed in two possible places of the fracture initiation (near the outer surfaces of the specimens and at the center of their cross sections, where hydrogen effect is possible). The key role of nonmetallic inclusions was noted in the destruction process of all puddled steels in artificial acid rain (Fig. 3.14). As a result of local electrochemical corrosion along the interfaces of the ferrite matrix with nonmetallic inclusions, deep damages formed around them. Such damages usually began to form from the outer surface of the specimens, and sometimes they extended to the whole cross sections (Fig. 3.14a). Bridges between long delaminations along the interfaces of the ferrite matrix with inclusions destroyed by their stretching up to critical state, with the following fracture by shear (Fig. 3.14b), or brittle transgranular cleavage (Fig. 3.14c). The signs of local corrosion with fine traces of electrochemical interaction on practically smooth facets of the fracture surfaces were also observed (Fig. 3.14d). Such smooth facets were considered as the surfaces originated by the delamination of nonmetallic inclusions from the matrix.

Analysis of the central part of smooth specimen's fracture surfaces from puddled steels after SCC test in artificial acid rain practically did not reveal non-metallic inclusions in the bottom of large and deep holes initiated by nonmetallic inclusions

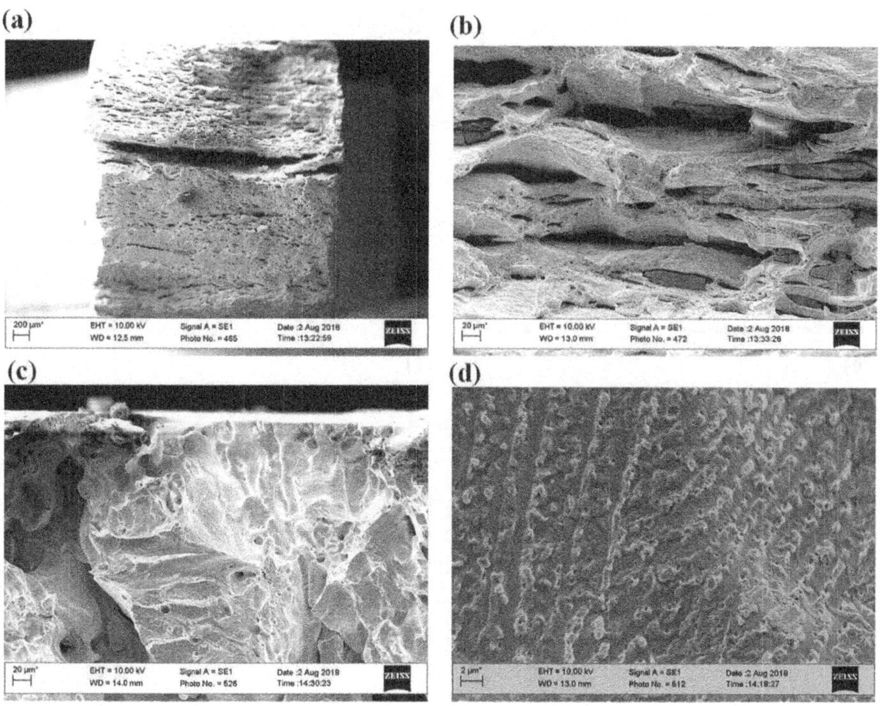

Fig. 3.14 Fractograms **a, b** of the steel (2) from the Sand Bridge element and **c, d** the steel (1) from the element of the Main Railway Station obtained in the vicinity of the outer surface of smooth specimens after SCC test in artificial acid rain

(Fig. 3.15). In contrast, most inclusions could be observed at the bottom of the holes on the fracture surfaces of specimens of the same steels tested in air. This is the first and the main fractographic feature of the corrosion environment effect on all puddled steels. Nonmetallic inclusions completely lost cohesion with the ferrite matrix under the influence of corrosive environment, thereby, they could not hold in the locations which they had occupied before the tests. As a result, such inclusions are separated

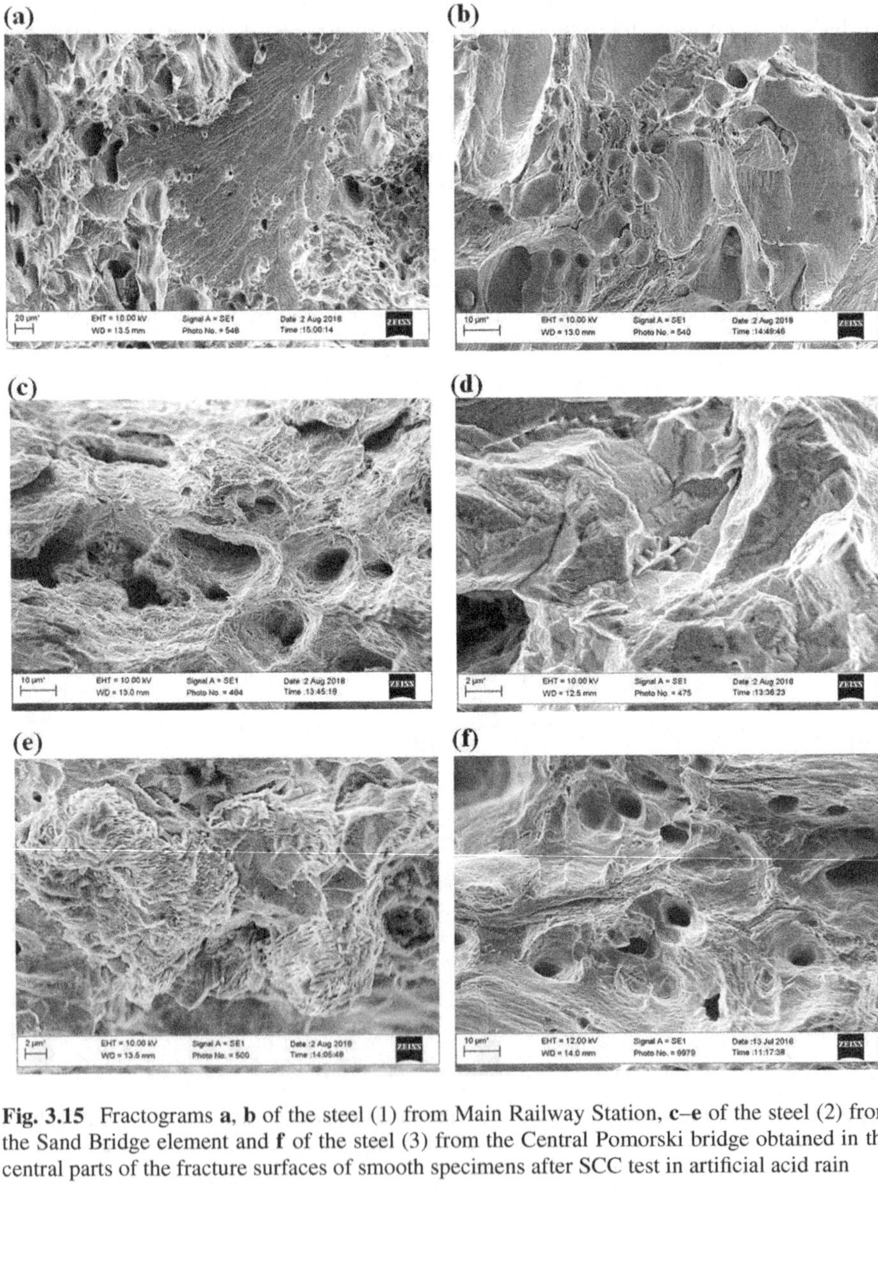

Fig. 3.15 Fractograms **a**, **b** of the steel (1) from Main Railway Station, **c**–**e** of the steel (2) from the Sand Bridge element and **f** of the steel (3) from the Central Pomorski bridge obtained in the central parts of the fracture surfaces of smooth specimens after SCC test in artificial acid rain

from both parts of the fracture surfaces during the final destruction of the specimens. Inclusions decohesion from the matrix during the testing of puddled steels in corrosive environment is quite possible due to favorable conditions for electrochemical corrosion on the microstructural level. The main components of puddled steels are ferrite, inclusions of cementite and oxides. Inclusions have more positive electrode potential than ferrite. The contact of the steels with heterogeneity at the microlevel with an electrolyte solution leads to the appearance on their surface of numerous galvanic short-circuited elements with nonmetallic inclusions as cathodes and ferritic regions as anodes. As a result, an active work of microgalvanic couples becomes possible and ferrite component of structure dissolves (especially actively along the interfaces with nonmetallic inclusions). Thus only large and deep holes with a diameter of 5...30 μm in the previous locations of inclusions remain on the fracture surfaces of specimens (Fig. 3.15c, f). Similar pits were revealed during in situ observation of the fracture of the Duplex Steels under the influence of artificial seawater. It was shown that the pits initiated from the inclusions of various composition (Si, Al, Mg, Ca, Ti, Mn and S) with further dissolution of the ferrite matrix [21].

Moreover quite large areas of transgranular fracture (with the typical cleavage facets) were revealed on the fracture surface of the steel (1) from the Main Railway Station. It is consistent with minimal values of the plasticity characteristics for this steel compared to other puddled steels under SCC test (Fig. 3.11). These areas were located along the bridges between parallel grooves formed by the neighboring delaminations along the boundaries between the chains of nonmetallic inclusions and the ferrite matrix (Fig. 3.15a).

These cleavage areas were observed practically over the whole fracture surface of the specimen on the background of ductile fracture with typical dimples relief. It was a main feature of the failure of the steel (1) from the element of the Main Railway Station tested in corrosive environment. It was suggested that this feature was concerned with hydrogenation of degraded metal. This conclusion agrees with the data known from the literature that hydrogen could promote steels degradation during their long-term operation [22]. It is also known that steels can be further embrittled under the influence of hydrogen [23, 24]. According to the tensile test of the API X65 steel in hydrogen gas at a pressure of 20 MPa directly in the microscope column (in situ), brittle fracture by quasi cleavage was realized in hydrogen, whereas a ductile fracture mechanism by dimples formation was predominated at the test in nitrogen or in air [25]. Consequently under long-term operation of steels in hydrogenating environment, the effect of hydrogen can be manifested both by hydrogen embrittlement of steels and by intensification of their degradation. But in both cases, hydrogen embrittlement will be manifested by cleavage elements on the fracture surfaces of the specimens.

A distinctive fractographic feature of the steel (2) from the Sand Bridge element is corrosion cracking along the boundaries of subgrains in ferrite (Fig. 3.15d). These elements were revealed in the central part of the fracture surface in the vicinity of the crack tip formed by both banks of deep delaminations. It was assumed that such fractography elements could be caused by hydrogen, accumulated in the cavities formed between delamination banks. This hydrogen could be weakened cohesion

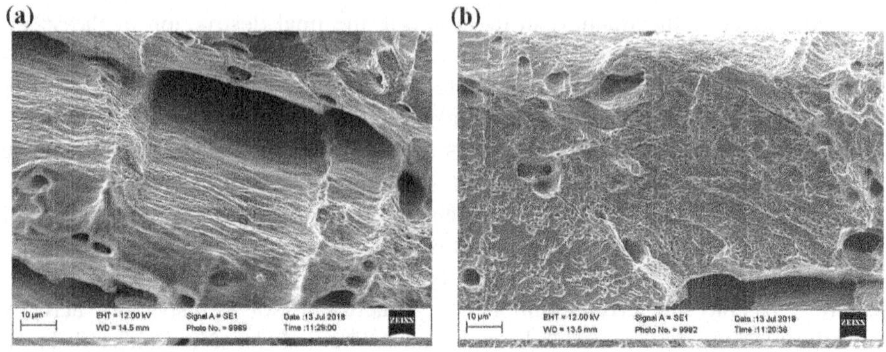

Fig. 3.16 Fractograms of the steel (3) from the Central Pomorski Bridge obtained in the central parts of the fracture surfaces of smooth specimens after SCC test in artificial acid rain

along the subgrain boundaries in the stress concentration zones in the vicinity of crack tips and facilitated the fracture along these boundaries. Under the influence of the same hydrogen accumulated in the delamination cavities, the bond between cementite and ferrite lamellas within the pearlite grains can also become weaker. As a result, an evident pearlite structure was visualized at the fracture surface (Fig. 3.15e).

Finally, on the fracture surface of the steel (3) from the Pomor Sea Central Pompey element with the maximum carbon content, a mechanism of ductile failure of the specimen after SCC test was discovered. Deep holes forming around nonmetallic inclusions (Fig. 3.15e), intensive deformation of bridges between them due to shear (Fig. 3.16a) or practically flat surfaces of delaminations (Fig. 3.16b), which also associated with anodic dissolution of the ferrite matrix in the vicinity of the inclusions were revealed as the main fracture features of this steel under the influence of environment.

In any case, the common feature for all puddled steels is the complete loss of cohesion of nonmetallic inclusions with the matrix, as evidenced by the almost complete disappearance of inclusions from their locations prior to testing in a corrosive environment.

At the same time, on the fracture surface of a specimen of the modern steel (4) (such as steel St3) tested for stress-corrosion cracking, the significant differences of the fractographic details compared to obtained at the tests in air were no revealed. But even in the case of the no operated steel the number of inclusions retaining the connection with the fracture surface after finishing of the stress corrosion cracking was less than at the test in air. The difference between their numbers on the fracture surfaces of the specimens obtained at the testing in air and in the acid rain environment did not exceed 15%.

3.3.9 Electrochemical Investigation of Steels

Corrosive aggressiveness of the atmosphere is characterized by climatic factors (temperature and relative humidity) and pollution in the form of acid-forming gases (CO_2, SO_2, NO, NO_2, H_2S and others), and fogs, and the content of dusty particles in the form of aerosols sprayed in the air. Depending on the relative humidity of air, the following types of corrosive atmosphere are distinguished: dry ($W < 60\%$); normal ($W = 61\ldots0.75\%$); humidity ($W > 75\%$). The mechanism of its action depends on the thickness of the moisture film on the surface of the structure. Dry atmospheric corrosion is realized at the thickness of the moisture film does less than of 10 nm. In normal relative humidity of air, the thickness of the moisture film does not exceed 0.1 μm, and in wet one—varies from 0.1 to 1 μm. The relative humidity of the air changes greatly even during one day and even more significantly during the year. Typically low and medium carbon steels for general use have a high resistance to atmospheric corrosion (under the influence of moisture with a neutral pH6...pH8) [26]. However, the acid rains (due to atmospheric pollutions) are common in industrial regions and significantly act on the corrosion resistance of these steels. The value of pH2 is characteristic of acid-containing precipitations, especially in the virgin period of snow or rainfall [27]. In addition, due to an acidification of the condensed moisture, the conditions arise for the development of crevice corrosion and corrosion under the exfoliated protective coating. Therefore, corrosion tests were carried out in slightly acidic environment with pH3 and pH4.4 (the chemical composition of used environment is given in the methodical part of this chapter).

Corrosion resistance of the investigated steels was estimated by the value of the polarization resistance R_p, which is the inverse of the corrosion current density j_{cor}, and, accordingly, the corrosion rate [26]. In the first approximation the corrosion activities of the puddled steels in the synthetic environments of acid rain at pH 4.4 and pH3 can be identified by analyzing the location of the polarization curves obtained on the specimens with polished surface (Fig. 3.17).

In particular, the corrosion potential E_{cor} of the puddled steel (2) from the Sand Bridge element is ~0.54 V compared to ~0.57 V for its prototype (modern steel (5) of the St3 type). Both analyzed steels had the comparable carbon content and the phase compositions with the ferrite-pearlite structures (Fig. 3.3). However, first of all these steels differed by the distribution of pearlite grains in the cross-section of the elements. The perlite grains were concentrated inside layers of the element cross section to the puddled steel, whereas its distribution throughout the element cross section to the modern steel was uniformly. In addition, a large number of nonmetallic inclusions oriented in the deformation direction of the structural elements during their manufacturing were observed in the puddled steel, but in modern one their number and dimensions were much smaller. Nevertheless, the most importantly, the puddled steel was operated for more than 100 years as the bridge element, whereas the modern one was in virgin state. Despite listed differences, the values of corrosion potentials of these steels not so much different.

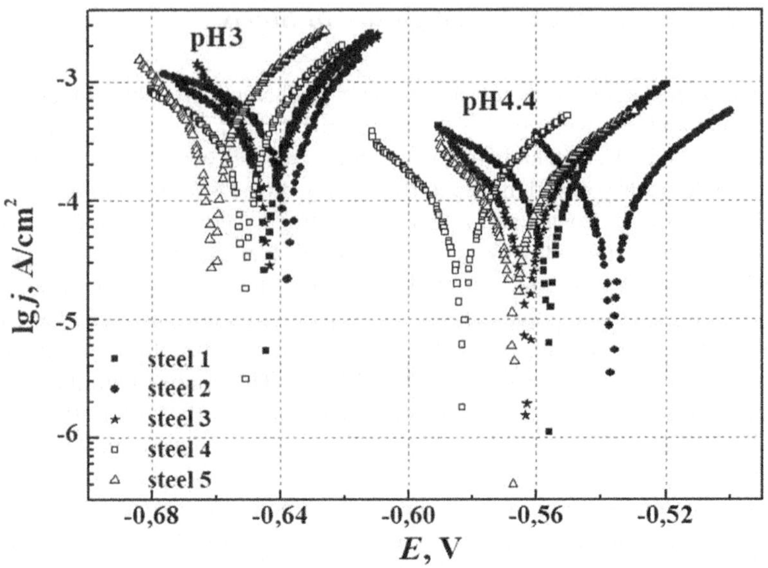

Fig. 3.17 The narrow sections of the potentiodynamic polarization curves (in the vicinity of the corrosion potential of E_{cor}) for the puddled (1, 2 and 3) and modern (4 and 5) steels obtained in the artificial environment of acid rain at the pH3 and pH4.4 (the pH values of the environment were varied by addition to the solution of the base composition the concentrated of H_2SO_4 acid). The polarization curves were obtained on the specimens with pre-polished surfaces

A qualitatively similar displacement was also observed when the corrosion potentials of steels with the lowest carbon content (modern steel (4) and puddled steel (1) from an element of the Main Railway Station) were compare. In this case, the E_{cor} level of the puddled steel (1) shifted to more positive value ~0.56 V compared to ~0.58 V for the modern steel.

It should also be noted that when the carbon content in the steels (and, correspondingly, the amount of perlite in their structures) increases, the corrosion potential of both modern and puddled steels shifts somewhat to the anode side. These results are consistent with the results obtained for the Fe-C alloys with different (over a wide range) contents of carbon (Fig. 3.18) [28].

This result was explained by the fact that when the carbon content in the steel increases, the fraction of the carbide/cementite phases in the structure of ferrite-pearlitic steels increases. The electrode potentials of such carbides are shifted relative to the potential of the ferrite matrix toward more positive values (to the anode region) [29]. Therefore, the polarization curves of steels with higher carbon content should also shift in this direction. However, the results obtained on the puddled steel (3) with the maximum carbon content did not confirm this regularity (Fig. 3.19).

The negligible difference between corrosion potential of the steel (3) of the Central Pomorski Bridge element and the E_{cor} value determined for the modern steel St3 was obtained. The polarization curve of steel (3) had shifted even to the side of cathode

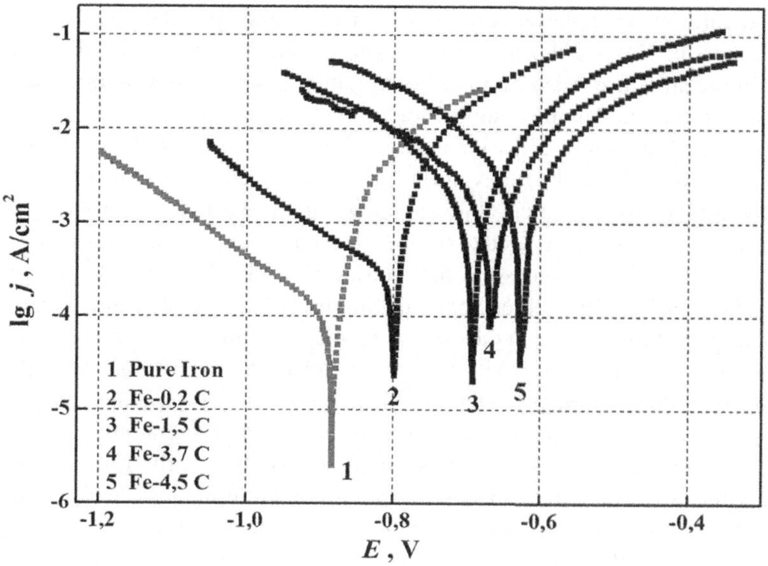

Fig. 3.18 Potentiodynamic polarization curves obtained for Fe-C alloys with different contents of carbon in naturally aerated 0.5 M H_2SO_4 system at a scan rate of 0.5 mV·c^{-1}. SSE—saturated sulfate electrode [28]

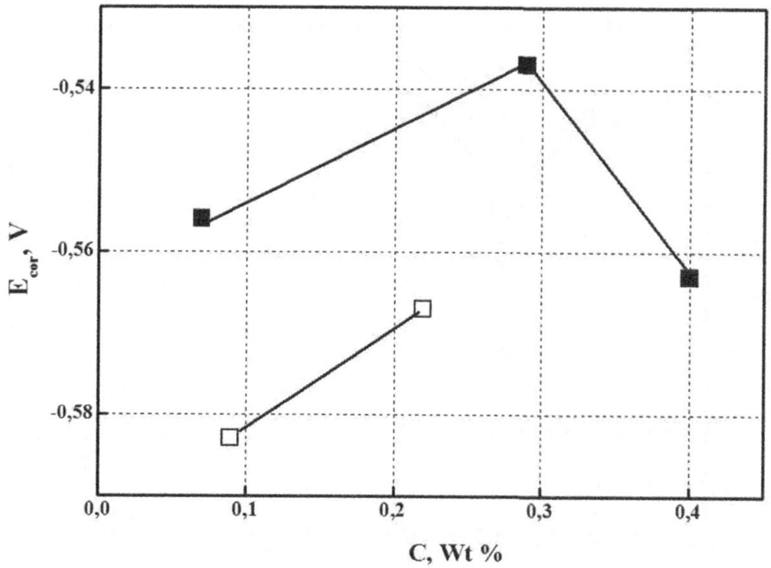

Fig. 3.19 Dependences of the corrosion potentials of E_{cor} versus of the carbon content of C in the puddled (1) and modern (2) steels obtained in the artificial environment of acid rain with pH4.4

region regarding the polarization curves for two other puddled steels with lower carbon content. Despite the maximum carbon content in this steel, the amount of perlite in the structure was minimal. Using a micro-X-ray spectral analysis, carbon was found in this steel in the composition of large (up to 1 μm in diameter) carbides. It was assumed that these inclusions had got into the steel from the cast iron.

The E_{cor} value of steel (3) deviation from established dependence E_{cor} – C may also be explained by a smaller range of changes in the carbon content in the puddled steels (~0.1…0.4% C) compared to the steel used in the mentioned study (0.2…1.5% C, Fig. 3.18) [28]. It is also possible that such a deviation can be caused by a significant difference in the content of other elements (Si, Mn, S, P) in puddled and modern steels.

Estimation of the change in the electrochemical parameter E_{cor} is more important in terms of its possible sensitivity this parameter to the operational degradation of these steels. Studies of the pipe steels of the main oil and gas pipelines, as well as the steam pipelines of the TPP, have shown that, due to their degradation during long-term operation the E_{cor} values of these steels are slightly shifted towards negative values of the potential with respect to the values inherent in these steels in the virgin states [30]. According to the results obtained on the steels from the old constructers, the E_{cor} values of the exploited puddled steels (1–3) shifted in an opposite direction compare to the modern no operated steels (4 and 5). Moreover, this trend was observed in both used environments (at pH4.4 and pH3, Fig. 3.20). This discrepancy can be explained by the predominant role of large slag inclusions detected in the

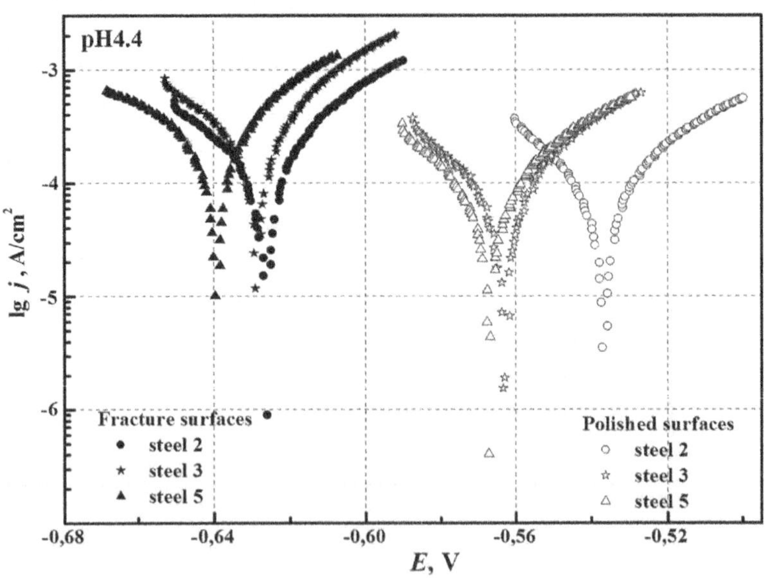

Fig. 3.20 The parts of the potentiodynamic polarization curves (in the vicinity of the potential of corrosion E_{cor}) for the puddled (2 and 3) and modern (5) steels obtained in the synthetic environment of acid rain at the pH4.4. These curves were obtained on the pre-polished surfaces (from right side) and on the fracture surfaces (from left side)

structure of puddled steels, but practically absent in their modern prototypes. As already described above, the big inclusions of iron oxides (with concomitant impurities of sulfur, phosphorus, manganese) were observed in all puddled steels (Fig. 3.2). In addition, large inclusions of iron carbides were detected in the structure of puddled steels (first of all in the steel (3) from the elements of the Central Pomorski Bridge).

The electrode potentials of slag inclusions, as like of the carbides are usually shifted towards more positive values relative to the potential of the ferrite matrix [29]. Thus, the favorable conditions for establishing local galvanic pairs between them are existed. It was assumed that, due to the structural galvanopairs in the puddled steels, the anodic polarization of the ferrite and, accordingly, the shift of the E_{cor} to the anode region will be more significant compared to modern steels. However, these differences were also insignificant (especially for steels 1, 3, 5, Fig. 3.19).

The destruction of the structural elements is usually occurred along the most damaged zones in their cross-sections. It was supposed that the polarization curves obtained on the fracture surfaces of the impact specimens will be more correctly for estimation of the technical state of the exploited steels than the curves obtained on the corresponding specimens with polished surfaces (usually used in classical electrochemistry). The polished surfaces of the traditional specimens for the electrochemical studies are usually prepared in arbitrary cross sections of steels. As a result the number of defects arisen during the long-term operation of the puddled steels has not necessarily been the greatest in these arbitrary cross-sections. At the same time the damages number in the puddled steels caused by their long-term operation must be the maximum on fracture surface of specimens (since the destruction of metal occurs exclusively along the least energy-consuming trajectory). It would be hoped the specimens for electrochemical studies in the form of fracture surfaces of impact specimens would show a higher sensitivity to the damage of exploited steels. This approach was first tried and showed a good sensitivity to the change of the technical state of the steels caused by their long-term operation on the main gas pipelines [31].

The obtained results have confirmed the higher corrosion activity of the fracture surfaces of impact specimens in comparison with polished surfaces (it is sufficient to compare the values of the corrosion potentials of the corresponding specimens, Fig. 3.20). The polarization curves of analyzed steels obtained on the fracture surfaces of impact specimens in environment with pH4.4 have shifted to the cathode region (side of more negative potentials) relative to the corresponding curves obtained on the polished specimens.

The order of the curve's disposition relative to each other has not changed compared to the tests on polished samples and the curve of modern steel (5) has taken the extreme left position (Fig. 3.20). However, the curve of puddled steel (2) has shifted to the anode region. Consequently, a relatively large discrepancy between the position of the polarization curves for the puddle steel (2) (with a ferrite-pearlite structure) and for the steel (3) (with carbides against a background of a practically ferrite matrix), obtained on the polished samples, had been practically disappeared using fracture surfaces of impact specimens of the same steels. This is consistent with the conclusion of fractography analysis testified the almost similar fracture mechanisms for both steels. Namely, in the spontaneous failure zones of both puddle steels the brittle fracture with typical cleavage elements was observed. The similar ductile fracture with formation of dimples was typical in the vicinity of stress concentrators of these steels. Deep delaminations along the slag inclusions, oriented in the direction of destruction and extended deep below the surface of fracture, were also observed on the both parts of fracture surfaces (with ductile and cleavage mechanisms). It was assumed (considering all the foregoing) that the absence of influence of different carbon content in steels (2) and (3) on their corrosion potential values, determined on the fracture surfaces of impact specimens, can be explained by the identity of their failure mechanisms.

To assess the changes in the technical state of the steels under the influence of various factors, the polarization resistance R_p is also successfully used [32]. It is one of the electrochemical characteristic considered as the most sensitive to degradation. The value of R_p is inversely proportional to j_{cor}, characterizing the corrosion resistance of steel. The higher the polarization resistance of the steel, the higher its corrosion resistance is. Analyzing the R_p values obtained after electrochemical tests of polished specimens in corrosive environments with different pH levels it is possible to conclude that decrease of the pH level leads to a significant decrease of the R_p values (Fig. 3.21).

When, the carbon contents in steels increases, the polarization resistance of both puddled and modern steels decreases (Fig. 3.22). A similar decrease in R_p is observed irrespective of the pH of the environment.

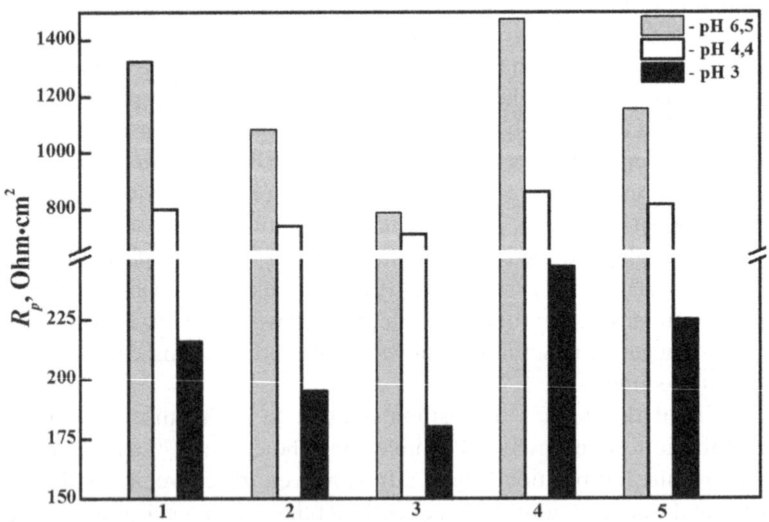

Fig. 3.21 The polarization resistance values R_p of the puddled steels (1, 2 and 3) and modern steels (4 and 5), obtained in an environment of 3% aqueous NaCl solution with pH6 (*grey columns*, [33]) and in the environments of the artificial acid rain with pH4.4 (*white columns*) and pH3 (*black columns*) using the traditional specimens with previously polished surfaces

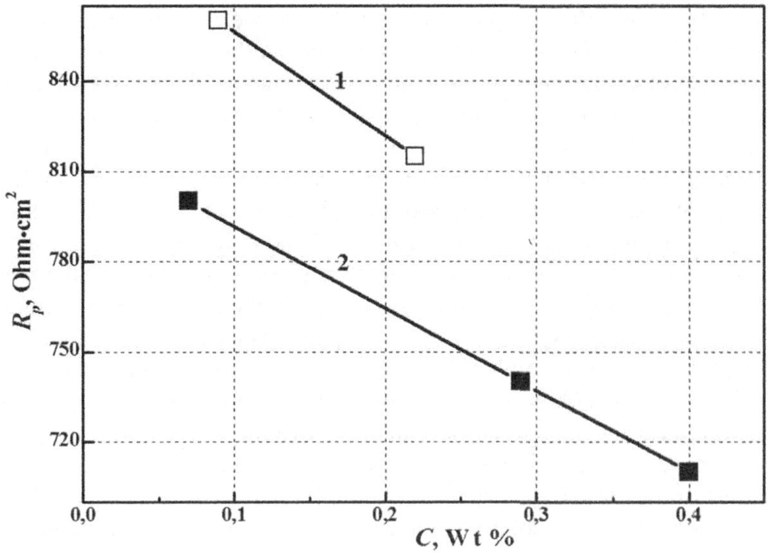

Fig. 3.22 The change in the polarization resistance R_p versus the carbon content in modern (1) and puddled (2) steels tested in a synthetic environment of the acidic rain at pH4.4 with using of polished samples

It is known that the corrosion rate of the relative pure iron in neutral environment [26] and in acid solutions are much less than corrosion rate of iron or steel with a high content of such elements as carbon, nitrogen, sulfur and phosphorus [28]. It has already been mentioned above that the electrode potentials of slag inclusions and carbides are shifted toward positive values with respect to the potential of the ferrite matrix. Thus, the favorable conditions for the set formation of local galvanic couples between them have already existed in puddled steels from the beginning of their exploitation. The anodic process of iron dissolution was localized on an anodically polarized ferrite, and the cathode one—on the inclusions. Therefore, the more inclusions in the steel structure, the more favorable conditions are created for local (at the structural level) electrochemical corrosion.

Galvanic corrosion is widely regarded as a type of local electrochemical corrosion and occurs when two dissimilar conductors are in electrical contact with one another due to a common conductive electrolyte. A very clear example of galvanic corrosion was discovered in the 1990s during the planned survey of the Statue of Liberty. Galvanic corrosion was detected between the outer copper shell and the wrought iron of structural support elements [27]. This is an example of galvanic corrosion at the macro level. At the same time, special attention has recently been paid to local aspects of galvanic corrosion at the microstructural level, which can accelerate general, point and ulcer corrosion.

The results of studying the corrosion process of four Fe-xC alloys with different contents of ferrite and cementite (or graphite) are shown the increase of these alloys corrosion rate in the sulfuric acid solution with increase of carbon content in them. The decrease of the average distance between the pearlite lamellae in the analyzed alloys, the rate of corrosion reactions appreciably increased [28]. These results can be easily explained from the standpoint of the galvanic corrosion model, since the anodic polarization of the ferrite phase increases with decreasing average distance between the pearlite lamellae. After all, cementite serves as a conductor for electrons on which hydrogen ions are recovered [34]. The cathode regions with a lower overvoltage generally arise on the cementite inclusions. The discharge of hydrogen ions and facilitating the cathode depolarization reaction, which limiting the corrosion rate process occur on these inclusions. As a result, the formation of micro-galvanic elements between the cementite and the ferrite phase and acceleration of the selective dissolution of ferrite and the influence on the kinetics of the galvanic effect take place [35, 36]. The effect of cathode regions associated with cementite becomes significant in steels with a carbon content of more than 0.2 wt% [37].

It is also known that if cementite exists as a separate phase, the effect of the carbon content on the corrosion resistance of steel in HCl solutions becomes stronger [38]. In this case, the continuous dissolution of ferrite leads to an increase in the area occupied by the cementite. As a result, the number of places favorable for the adsorption reaction of corrosion-aggressive components of the environment and, accordingly, the galvanic dissolution of the ferrite phase is constantly increasing [28].

In carbon steels, cementite can additionally adversely affect the course of galvanic reactions due to the complication of the protective films formation [39]. The corrosion rate of carbon steels increases in a neutral deoxygenated 5% NaCl solution saturated with CO_2 (pH6) when the carbon content increases. An adhesion of the passive films on the surface depends on the morphology of cementite (lamellar, globular, etc.) [40]. While the corrosion resistance of low-alloy steels in a Cl^- containing medium increases with decreasing carbon content [41].

Analyzing the results of electrochemical studies obtained under identical conditions, the R_p values of all puddled steels were always lower than for the corresponding modern steels (Fig. 3.23). The ratios of the polarization resistances R_p^{op}/R_p^{virg} for all puddled steels have always been less than one, corresponding to their modern prototypes. It indicates that the polarization resistance, determined on specimens with a polished surface, can be a sufficiently informative electrochemical indicator of the technical state of steels. The gradation of puddled steels after long-term operation using this parameter can be carried out by estimation of their sensitivity to the effects of corrosive environment in comparison with modern steels of appropriate chemical composition. However, the sensitivity of this indicator to the steels degradation does not exceed 15%. Nevertheless, it can be concluded that puddled steel (1) with the lowest carbon content is least affected by operational degradation. However, unfortunately, this conclusion does not agree with the results of mechanical tests. According to the impact test, a catastrophically low value of toughness (0.06 kJ /m^2) was determined for this steel with a practically ferrite structure after its long operation in the Main Railway Station elements (Table 3.2).

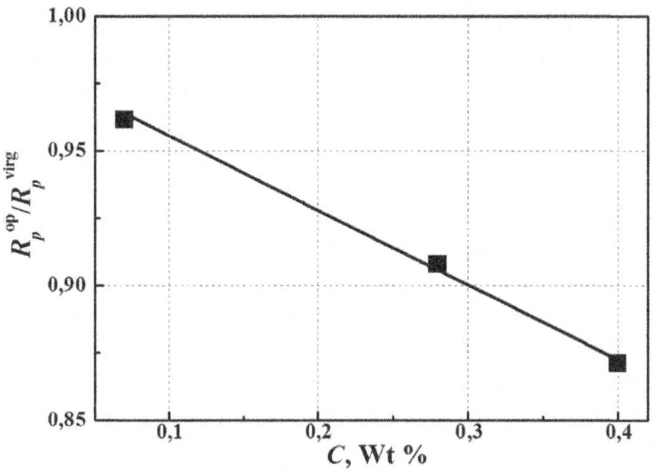

Fig. 3.23 Estimation of the operational degradation of puddled steels depending on the content of carbon C using the ratio of the polarization resistances of the puddled steels R_p^{op} and modern steels with the corresponding chemical composition R_p^{virg}, obtained on polished specimens in the artificial environment of acid rain with a pH4.4

The values of the polarization resistance R_p^* of all studied steels, obtained in environment with pH4.4 on non-traditional specimens (fracture surfaces of impact specimens), were significantly lower than corresponding R_p values obtained on traditional polished samples (Fig. 3.24a in comparison with Fig. 3.22). At the same time, polarization resistance values of modern steels are higher in comparison with puddled ones.

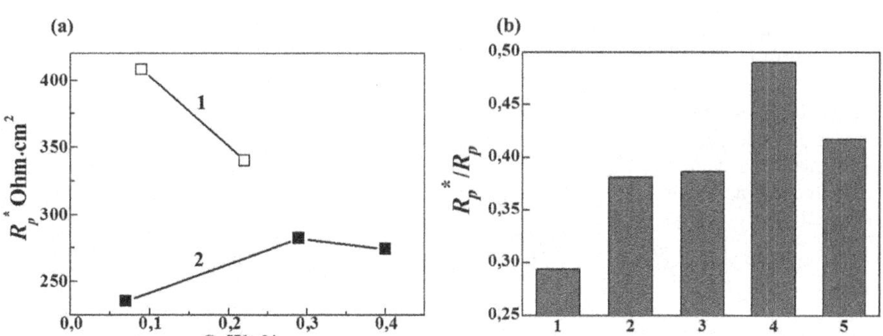

Fig. 3.24 **a** The values of the polarization resistance R_p^* for modern (1) and puddled (2) steels obtained in artificial environments of acid rains with pH4.4 using in the electrochemical studies the non-traditional specimens—fracture surfaces of impact samples and **b** the ratios between the polarization resistance values of these steels obtained on the specimens with a fracture surfaces R_p^* relative to values obtained using the traditional specimens R_p

Even for modern steels, a significant reduction of the polarization resistance (by almost 50%) was obtained when polarization resistance value determined by using the fracture surfaces of impact samples (Fig. 3.24b). However, the maximum reduction of the R_p^* was observed in puddled steel (1). It was mentioned above that the destruction of steels takes place exclusively along the trajectory with the least energy consumption. It is clear that the amount of macro—and microdamages should be along the fracture surface maximum. Accumulation of macrodamages could be caused by the presence in the puddled steels large slag inclusions. Their connection with ferrite matrix could be disturbed during the long operation of the structural elements, in particular, due to accumulation of hydrogen inside them. It is known that hydrogenation of steels is possible even in conditions of atmospheric corrosion [42, 43]. In particular, hydrogen can penetrate into the metal through local changes in pH and potential near the corroding surface due to the evaporation of a thin film of water. It is also reported that as a result of electrochemical corrosion of the metal surface, hydrogenation of the entire volume of metal becomes possible [44]. Subsequently, hydrogen can accumulate along the interphase boundaries of the inclusions and the matrix. This hydrogen, on the one hand, facilitates the separation of inclusions from the matrix. On the other hand, hydrogen reduces the ability of structural elements with such internal damages to resist external influences of the loadings and corrosive environments.

The accumulation of micro damages in the vicinity of fracture surfaces was also associated with a very high degree of metal plastic deformation proceeding to the

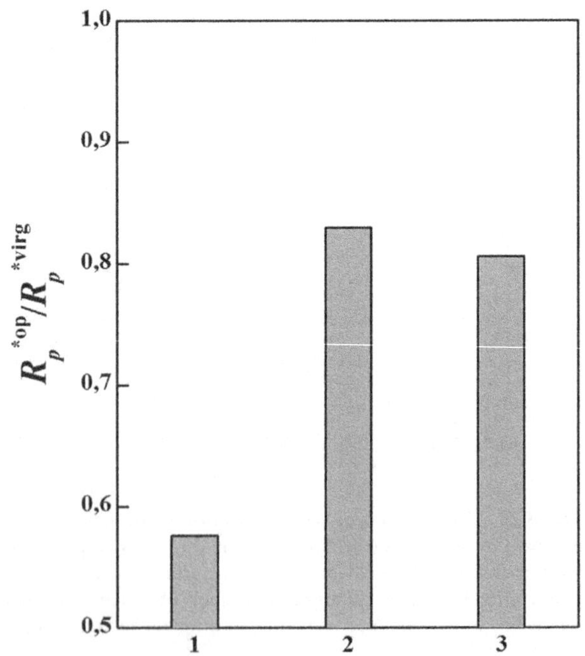

Fig. 3.25 Estimation of the operational degradation of puddled steels (1, 2, 3) using the ratios of the polarization resistances of the puddled steels R_p^{*op} and modern steels with the corresponding chemical composition R_p^{*virg}, obtained on the fracture surfaces of the impact specimens in the synthetic environment of acid rain with a pH4.4

final destruction of the specimens. After all, the dislocations density and the number of dislocation sources increases by orders of magnitude due to plastic deformation of specimens [45]. Dislocation pileups and slip lines have a low electrochemical potential. Therefore, the places of their locations are subject to intensive local anodic dissolution. On the other hand, defects introduced by cold processing and connected with carbon atoms (Cottrell's atmosphere) are characterized by low hydrogen over voltage. It promotes the process of cathodes depolarization. In the absence of carbon the apparent hydrogen over voltage of areas with imperfections is the same as in iron, and their tendency to anomalous dissolution is not manifested [46]. Thus, the superposition of the effects of facilitating of both anodic and cathode reactions of electrochemical corrosion is possible on such defects as dislocation pileups and slip lines.

The usage of local scanning methods (scanning the kelvin probe and local investigation of the electrochemical impedance) made possible to show that the residual stresses also reduce the ability of the steel to the surface passivation in the electrolyte and further corrosion acceleration. The mentioned above factors could be cause by such a significant decrease of the R_p^* value estimated on the fracture surfaces of the impact specimens as compared to the polarization resistance on the polished surfaces.

The effect of operational degradation of puddled steels in comparison with modern one with corresponding composition using the relative changes of the values of their

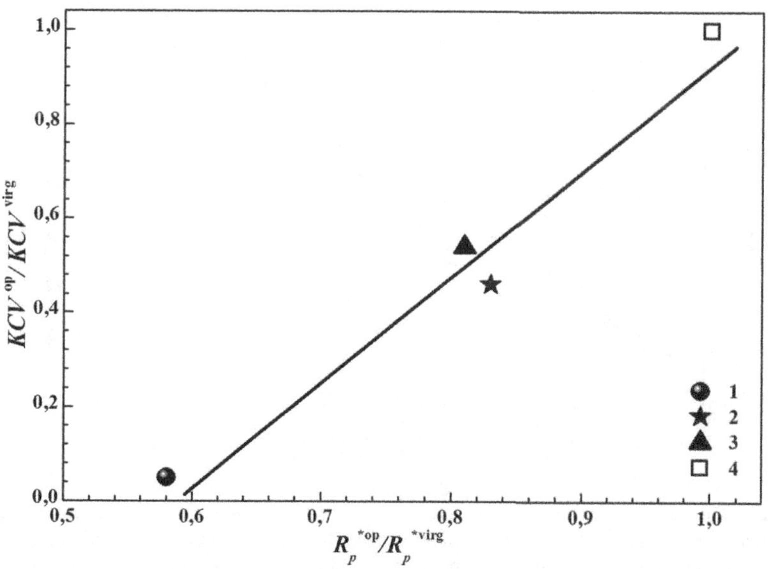

Fig. 3.26 The relationship between the ratios of the KCV^{op}/KCV^{virg} and R_p^{*op}/R_p^{*virg} for puddled steels after their long-term operation on the various objects (Main railway station (1), Sand Bridge (2) and Central Pomorski Bridge (3)). An electrochemical investigation was carried out in the synthetic environment of acid rain with pH4.4

polarization resistance was much more pronounced on the specimens in the form of fracture surfaces (Fig. 3.25 in comparison with Fig. 3.23).

In addition, the relative decrease of the R_p* value (obtained on the fracture surfaces of impact specimens) caused by an operational degradation of steel (1) with a minimum carbon content was the highest (more than 40%) regarding to the values obtained for other puddled steels (Fig. 3.25). While the decline of this ratio for puddled steel (1) was the lowest on the samples with polished surfaces (Fig. 3.23). The results of the polarization resistance values estimation of all investigated steels by using non-traditional specimens with the fracture surfaces are in a good agreement with the results of the impact tests (Fig. 3.26). Thus, the usage of such an electrochemical characteristic as the relative decrease in the polarization resistance due to steel degradation caused by their long-term operation in the old constructions opens a good prospect for diagnostics of the current technical state of puddled steels.

Earlier [30, 32], on the base of results of mechanical and electrochemical studies of different carbon and low-alloy ferritic-pearlitic steels of various applications a dependence was got which useful for the diagnostics of their operational degradation (Fig. 3.27).

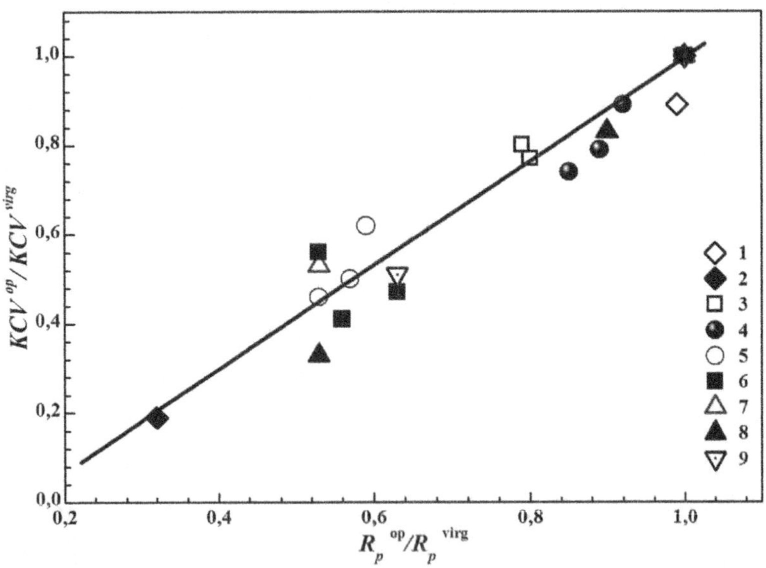

Fig. 3.27 The dependence between the ratios KCV^{op}/KCV^{virg} and R_p^{op}/R_p^{virg} for carbon and low-alloy ferritic-pearlitic steels of various objects of long-term operation [37, 42]: the St3 steel for the elements of water tower (1), and of lighthouse (2), and for stack discharge conveyor (3), the St3sp steel for the gantry cranes (4), and for the storage tanks for raw oil (5) and for the commercial oil (6), the 10GS steel of the main oil pipeline (7), the 12X1MF steel for the bending of a steam pipeline of the TPP (8), the 25X1M1F steel for the roller of the continuous casting machine (9), tested in 0.3 (1–8) and 3% (9) of the roll of the continuous casting machine (8) tested in 0,3 (1–8) and 3% (9) water solutions of NaCl

The established correlation dependence enables to predict the resistance to brittle fracture of such steels by changing their electrochemical characteristics and is the basis of a nondestructive method for diagnostics of the operational degradation of steels of this type. Unfortunately, the points obtained at the results of the SSC under SSRT for the puddled steels in the artificial acid rain environment (the linear dependence obtained in the first approximation on the basis of these tests is shown in Fig. 3.26) do not superimposed into the linear dependence in Fig. 3.27, which was obtained after the results of the tests in an aqueous solution of sodium chloride. It is possible that during tests in the artificial environment of acid rain the unaccounted factors of the corrosion influence were arisen that requires deeper investigations of these features.

References

1. Lesiuk G, Szata M (2011) Aspects of structural degradation in steels of old bridges by means of fatigue crack propagation. Mater Sci 47(1):82–88
2. Lesiuk G, Szata M (2010) Wpływ procesów degradacji mikrostruktur na udarność stali mostowych z przełomu XIX i XX w. Logistyka 6:1883–1892
3. Lesiuk G, Szata M (2010) Degradacja mikrostruktur elementyw konstrukcyjnych pochodzących z mostów wzniesionych na przełomie XIX i XX wieku. Zeszyty Naukowe–Wyższa Szkoła Oficerska Wojsk Lądowych im. gen. T. Kościuszki 42(1):96–109
4. Bowman MD, Piskorowski AM (2004) Evolution and repair of wrought iron and steel structures in Indiana. Final report FHWA/IN/JTRP–2004/4. Purdue University, West Lafayette, Indiana
5. Flemings MC, Ragone DV (2009) Puddled: a new look at an old process. ISIJ Int 49(12):1960–1966
6. Samuels LE (2010) Light microscopy of carbon steels. ASM International, OH, USA
7. Lesiuk G, Szata M, Bocian M (2015) The mechanical properties and the microstructural degradation effect in an old low carbon steels after 100-years operating time. Arch Civ Mech Eng 15(4):786–797
8. Jesus AMP, Silva ALL, Figueiredo MV, Correia JAFO, Riberio AS, Fernandes AA (2011) Strain-life and crack propagation fatigue data from several Portuguese old metallic riveted bridges. Eng Fail Anal 18(1):148–163
9. Lesiuk G, Correia JAFO, Jesus AMP, Kucharski P (2016) Fatigue crack propagation behavior of old puddle iron including crack closure effects. 21st European Conference on Fracture, ECF21, vol 2. Elsevier, NY, pp 3218–3225
10. Lesiuk G, Szata M (2014) Fatigue properties and fatigue crack growth in puddle steel with consideration of microstructural degradation processes after 100-years operating time. XVII International Colloquium on Mechanical Fatigue of Metals, ICMFM-17, vol 74. Curran Associates, NY, pp 64–67
11. Ostash OP, Vol'demarov OV, Hladysh PV (2012) Cyclic crack resistance of the steels of bends of steam pipelines after long-term operation. Mater Sci 48(4):427–437
12. Nykyforchyn HM, Kutnyi AO, Student OZ, Krechkovs'ka HV, Zvirko OI, Kurnat IM (2014) Structure and properties of the steels of hyperboloid gridshell Shukhov's towers after long-term operation. Mater Sci 49(6):787–795
13. Megaencyclopedia of Cyril and Methodius. http://megabook.ru/article/. Accessed 29 Aug 2018
14. Salimi A, Zadeh HM, Toroghinejad MR, Asefi D, Ansaripour A (2013) Influence of sample direction on the impact toughness of the API-X42 microalloyed steel with a banded structure. Mater Technol 47(3):385–389

15. Kravchuk A, Kondryakov I, Panasenko O, Kharchenko V (2017) Influence of the cutting direction of the charpy specimens from the steel 10GN2MFA on the components of the fracture energy and the ductile-to-brittle transition temperature. Sci J Ternop Natl Tech Univ 88(4):26–34
16. Maisuradze MV, Ryzhkov MA (2017) The dependency of mechanical properties on the microstructure anisotropy index of some alloyed steels. AIP Conference Proceedings, vol 1915, no 1. AIP Publishing, Melville, pp 040037: 1–4
17. Nykyforchyn HM, Student OZ, Krechkovs'ka HV, Markov AD (2010) Evaluation of the influence of shutdowns of a technological process on changes in the in-service state of the metal of main steam pipelines of thermal power plants. Mater Sci 46(2):177–189
18. Student OZ, Krechkovs'ka HV (2012) Anisotropy of the mechanical properties of degraded 15KH1M1F steel after its operation in steam pipelines of thermal power plants. Mater Sci 47(5):590–597
19. Student OZ, Krechkovs'ka HV, Palashchuk TE, Hladkyi YaM (2018) Influence of the long-term operation of 12Kh1MF steel of the bends of main steam pipelines of thermal power plants on its mechanical properties. Mater Sci 53(4):460–467
20. Nykyforchyn HM, Student OZ, Markov AD (2007) Abnormal manifestation of the high-temperature degradation of the weld metal of a low-alloy steel welded joint. Mater Sci 43(1):77–84
21. Garfias L (2015) The effect of microstructure on the pitting resistance of duplex stainless steels. NACE–International Corrosion Conference Series
22. Nykyforchyn HM, Kurzydlowski K-J, Lunarska E (2008) Hydrogen degradation of steels under long-term in-service conditions. Environment-induced cracking of materials. Prediction, industrial developments and evaluations, vol 2. Elsevier, Amsterdam, pp 349–361
23. Panasyuk VV (2014) Decohesive concept of the interaction of hydrogen with metals. Mater Sci 50(2):161–169
24. Dmytrakh IM, Leshchak RL, Syrotyuk AM, Lutyts'kyi OL (2014) Influence of the bulk concentration of hydrogen in the metal on the specific features of deformation of low-alloy pipe steel. Mater Sci 50(2):170–178
25. Lee Y-H, Lee Hm, Kim Y (2011) Mechanical degradation of API X65 pipeline steel by exposure to hydrogen gas. Metals Mater Int 7(3):389–395
26. Shreir LL (eds) (1976) Corrosion. Metal. Environment Reactions, vol 1. Newnes, Butterworths, London, pp 1232
27. Lee H-Y (2008) Investigation on the use of iron and steel for restoration purposes during 19th and 20th century. Master's Thesis. Spain, Technical University of Catalonia, p 59
28. Ferhat M, Benchettara A, Amara1 SE, Najjar D (2014) Corrosion behaviour of Fe-C alloys in a Sulfuric Medium. J. Mater. Environ. Sci 5(4):1059–1068
29. Slag and its relation to the corrosion characteristics of ferrous metals–National Slag Association 172-13. p 23. http://www.nationalslag.org/sites/nationalslag/files/documents/nsa_172-13_slag_and_corrosion.pdf. Accessed 29 Aug 2018
30. Tsyrul'nyk OT (2014) Application of the electrochemical methods in the diagnostics of the engineering state of structural materials. Mater Sci 49(4):449–460
31. Nykyforchyn H, Tsyrulnyk O, Zvirko O (2018) Electrochemical fracture analysis of in-service natural gas pipeline steels. Structural Integrity Procedia
32. Zvirko OI (2017) Electrochemical methods for the evaluation of the degradation of structural steels intended for long-term operation. Mater Sci 52(4):588–594
33. Zvirko OI, Nykyforchyn HM, Szata M, Kutnyi A, Lesiuk G (2014) Corrosion degradation of old structures steels. Physicochem Mech Mater 10(1):29–33 (in Ukraine)
34. Matlosz M, Magaino S, Landolt D (1994) Electrochem Soc. 141:410–418
35. Epelboin I, Keddam M (1972) Electrochim Acta 17:177–186
36. Belkhaouda M, Bazzi L, Benlhachemi A, Salghi R, Hammout B, Kertit S (2006) Appl Surf Sci 252:7921–7925
37. Dugstad Lunde L, Videm K (1994) Corrosion/94, 14

38. AbdulAzim AA, Sanad SH (1972) Effects of acid concentration, C-content and temperature on the corrosion rate of steel in HCl. Corros Sci 12(4):313–324
39. Nešić S, Lunde L (1994) Carbon dioxide corrosion of carbon steel in two-phase flow. Corrosion 50(9):717–727
40. López DA, Schreiner WH, de Sánchez SR, Simison SN (2003) The influence of carbon steel microstructure on corrosion layers: An XPS and SEM characterization. Appl Surf Sci 207(1–4):69–85
41. Guo Jia, Yang Shanwu, Shang Chengjia, He Xinlai (2009) Influence of carbon content and microstructure on corrosion behaviour of low alloy steels in a Cl^-containing environment. Corros Sci 51(2):242–251
42. Omura T, Kudo T, Fujimoto S (2006) Environmental factors affecting hydrogen entry into high strength steel due to atmospheric corrosion. Mater Trans 47(12):2956–2962
43. Tsuru T, Huang Ya, Ali MdR, Nishikata A (2005) Hydrogen entry into steel during atmospheric corrosion process. Corros Sci 47(10):2431–2440
44. Nykyforchyn HM, Kutnyi AO, Student OZ, Krechkovs'ka HO, Zvirko OI, Kurnat IM (2014) Structure and properties of the steels of hyperboloid gridshell Shukhov's towers after long-term operation. Mater Sci 49(6):787–795
45. Nazarov A, Vivier V, Thierry D, Vucko F, Tribollet B (2017) Effect of mechanical stress on the properties of steel surfaces: scanning kelvin probe and local electrochemical impedance study. J Electrochem Soc 164(2):66–74
46. Foroulisand ZA, Uhlig HH (1964) Effect of cold-work on corrosion of iron and steel in hydrochloric acid. J Electrochem Soc 111(5):522–528

Chapter 4
Case Studies: Structural, Fractographic and Mechanical Aspects of the Steels Degradation of the Hyperboloid Gridshell Towers

Abstract The typical corrosion damages of structural elements of water tower design engineer V. Shukhov in Mykolayiv (Ukraine) were analyzed. The technical state of the old and repair (used for tower renovation) steels after operation during ~110 and 70 years respectively were evaluated. For their validation the metallographic and fractographic investigations and estimation of the mechanical properties such as hardness, strength, plasticity, impact toughness and stress corrosion cracking (SCC) at slow strain rate test (SSRT) were used. Low values of hardness and brittle fracture resistance were typical for both exploited steels. The mechanical characteristics of the old steel obtained in air were worsened compared with the repair one. Moreover, the plasticity characteristics of both steels, determined at SCC tests in synthetic acid rain environment, are lower than in the air. It was suggested that fractography elements had appeared due to hydrogenation effect of steels at the influence of the corrosion environment (the really acid rain) during their long-term operation in the structural elements of the water tower. The whole complex of studies had confirmed a degradation of steels' characteristics caused by the development of scattering damages during their operation.

4.1 Brief Information on Gridshell Designs Based on the Principle of One-Sheet Hyperboloid Surface

Preservation of the old and original metal engineering structures continues to attract attention in many countries of the world. Regarding Ukraine, it is primarily the gridshell hyperboloid towers, installed in the early twentieth century on the projects of engineer Vladimir Shukhov. In 1899, engineer V. Shukhov had been patented the principle of constructing hyperboloid gridshell structures based on a hyperboloid of revolution. It can be created by rotating a hyperbola around one of its main axes. A one-sheet hyperboloid is a connected surface having negative Gaussian curvature at each point. Through any point of this surface, two intersecting lines can be drawn, completely belonging to it. Thus the one-sheet hyperboloid surface can be formed by the set of straight lines. If the steel beams would be placed along these lines

G. Lesiuk et al., *Degradation Theory of Long Term Operated Materials and Structures*, Structural Integrity 15, https://doi.org/10.1007/978-3-030-43710-7_4

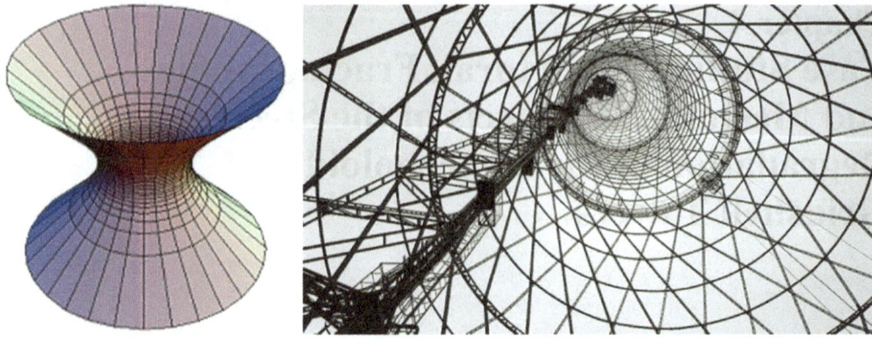

Fig. 4.1 The schematic representation of hyperboloid and the fragment of the gridshell structure of the Shukhov's tower

the shape of the structure could retained under the influence of external loading (if these elements are even connected using the hinges). In addition to direct intersecting beams made of angular rolled steel, in the Shukhov's towers the horizontal rims were used (Fig. 4.1). These rims were located at different levels along the height of the structures. In lowest level of the structure the rim was obtained by curving at a circle of the thick (up to 10 mm) strip from the rolling steel. The angular rolling steel was used on the higher levels of tower. The stability of the structure was ensured by numerous riveted connections of straight steel corners between each other and with horizontal rims in the places of their intersection [1]. During the installation of the towers the straight angles of the metal profile was somewhat deformed throughout the shelves width of the angular profiles of both intersecting elements in order to ensure the maximum contact between them in the places of their connection by rivets. Thus the elements of Shukhov's towers were subjected to a preliminary deformation in places of their connection by rivets during the installation of towers.

The main hazard for high-rise structures is the loads caused by wind gusts. The gridshell design of the Shukhov's Tower enabled to minimize their influence. Openwork design of these structures had ensured in sufficient strength, high stability and low metal consumption. With a comparable height of structures, consumption of metal per unit height of the Shukhov's Tower was three times less than of the Eiffel Tower in Paris. According to the project, Shukhov's tower in Polybino was supposed to have a height of 350 m with an estimated weight of 2,200 tons. While the Eiffel Tower, which is 324 m high, weighs about 7,300 tons. Due to this advantage such principle of structures designing is used to this day during the building of modern high-rise structures. In particular, the Kobe Port Tower in Japan [2] (Fig. 4.2a), and TV towers in Australia, Ještěd in Czechia, Guangzhou in China [3] (Fig. 4.2b), and Khan Shatir Shopping-Entertaining Center in Kazakhstan, Aspire Tower in Qatar [4] (рис. 4.2c), are the vivid examples using of the hyperboloid principle during engineering design of modern buildings.

(a) (b) (c)

Fig. 4.2 General view of the modern objects built by using the principle of hyperboloid designing: **a** Kobe Port Tower in Japan, **b** Kwangchow in China, **c** Aspire Tower in Qatar

The first water tower in Ukraine, designed by V. Shukhov, was built in Lisichansk in 1896 at the Lyubimov and Sol've soda factory. This tower was similar with a tower built for the world exhibition in Nizhny Novgorod (1896).

Among the Shukhov's towers in Ukraine, two lighthouses in the Dnipro's Lyman (so-called Adzhigol Lighthouses in the Dnipro river estuary) continue to operate until now (Fig. 4.3a, b). In addition the several of water towers are no longer used for water supply. Among them are the towers in Mykolaiv (Fig. 4.3c), in Cherkasy (Fig. 4.3d), in the Bila Tserkva, in Konotop and Pomoshna. However, despite to their current non-functionality is important to preserve them as a memory of the original engineering solution and as the monuments of an industrial architecture of the late 18th and early 19th centuries [5–7]. In general, all of these unique constructures had undergone a significant corrosive effect. Consequently, listed constructions need to examination and restoration. In general the 191 of the Shukhov's Towers (from known near 200 towers each of which was unique) have been irretrievably lost during of the 20th century and can be completely lost already in the 21st century [8]. Moreover, dismantling of these structures with their destruction is not related to natural disasters or military actions. These towers were destroyed because their continued use for water supply has become impractical. To use them for another purpose (in particular, as museum exhibits) large investments were necessary. Therefore, it is very important to preserve at least those of these objects that are not yet irretrievably lost.

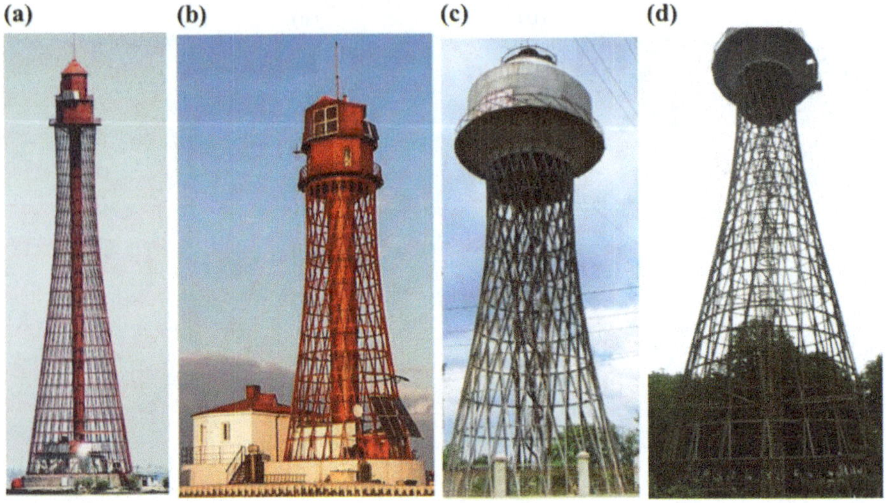

Fig. 4.3 General view of hyperboloid gridshell structures in Ukraine: **a** the Adzhigol lighthouse (1911) [9], **b** the small Stanislavsky lighthouse in the Kherson port (1911) [10], **c** water towers in Mykolayiv (1906), **d** hyperboloid towers in Cherkasy

The height of the Shukhov's tower in Mykolayiv is 32 meters (Fig. 4.3c). It is the oldest tower from existing in Ukraine. But the uniqueness of this tower is not only in its age, but also that the engineer V. Shukhov directly supervised the construction of the tower. This tower was the first in the world, which was used to supply the city by the water. In addition, unlike to other water towers, the old underground communications were preserved in Mykolayiv. Although today this tower is no longer used by its functional purpose, its location in the centre of the city and the preservation of the authenticity of most elements makes it an interesting landmark of industrial architecture. However, corrosion of metal can be an insurmountable obstacle to the preservation of this tower. Therefore, the results of the inspection of the state of tower elements and the assessments of the technical state of the metal will be useful in developing of means for the structures protection from corrosion and in calculating of its structural integrity, taking into account the change of the metal properties during its long-term operation.

The providing of the structural integrity of elements (which usually calculated with account of allowable stresses) complicated if the mechanical characteristics of the metal are degraded due to its long-term operation. In the 1950s, the towers were calculated in accordance with the limiting state. The conditions of the impossibility of further exploitation of the structures were determined with account of the loss of load capacity, critical damages, loss of stability and deformation of elements. After that, since the towers were no longer used with functional purposes such assessments were never repeated again.

It is important to know the non-supposed, but the actual state of the metal from these structures in order to preserve and plan the measures for their conservation and reconstruction. It is primarily concerned to estimation of the materials resistance to brittle fracture and to the mechanical characteristics during tensile test (especially at the SCC with SSRT in an artificial environment of acid rain), to analysis of the fractographic features of the degradation of old steels after more than a century of operation, and the steels used to towers reconstruction in the postwar period.

Some experience in assessing of the technical state of structural steels after long-term operation in elements of different objects is already accumulated. This primarily concerns to the assessment of the steels state used in the main oil and gas pipelines [11, 12], bridge and port facilities [13, 14], equipments of the TPPs [15–17] and petrochemical industry [18], vessels for storage of the petroleum products [19], etc. These objects are usually operated at the difficult conditions (at high loading and temperature and in aggressive environments).

The certain experience is also accumulated regarding the evaluation of metal state of old designs after its long-time operation [20–22]. The degradation signs in the surface layers and «in the bulk» of the metal were analyzed. The degradation of the surface layers of metal is associated with corrosive ulcers or pits and scratches or potholes due to corrosive or mechanical factors. Degradation "in the bulk" of the metal (or in a large part of thickness of the structural elements) leads to a decrease in its physical and mechanical properties. Structural-phase changes or scattered damages inside the metal in the form of micro- and nanoscale defects or microcracks can also testify about the degradation "in the bulk" of metals.

To connect the damaged elements of the Mykolayiv's tower during the repair and restoration works in the post-war period the welding technology was used instead the connection technology by rivets in the original design. The electrodes and welding technologies available at the time were used. Everything was aimed at the maximum rapid resumption of the functionality of the tower. At the same time, the welded joint becomes one of the weakest links in the welded structures after their long-time operation [23]. Thus, to ensure the integrity of the tower as a whole the inspection of the state of welded joints between the tower elements is especially significant.

4.2 Investigated Materials and Test Methods

The objects of study were the elements of the Shukhov's Tower in Mykolayiv fabricated from the rolled steels with angle profile after different operation time (Fig. 4.4). The tower consists from supporting elements made from the steel with angle profile and horizontally arranged rims curved from steel strips or the steel with angle profile. All the elements were installed in their respective positions and joined together using the rivets. During of the Second World War the tower in Mykolayiv was blown up (Fig. 4.4a) [24]. Its functionality was restored in 1944. The damaged elements were replaced by new one using welding technology for their connection between themselves and then tower was lifted. Today the tower remains in its place (Fig. 4.4b).

(a) (b)

Fig. 4.4 General view of the Shuhov's tower: **a** in postwar period [24]: **b** today

To analyze of the current state of tower's metal the authentic elements, connected by rivets, and the elements added during the repair work with using postwar welding technology were selected.

The chemical composition and mechanical characteristics, and the metallographic and fractographic features of the old (1906) and of the repair (1944) steels from the selected elements of water tower were analyzed.

The chemical composition of the degraded steels was evaluated using of the spark optical atomic-emission spectrometer SPECTROMAX LMF 0.5. The impact tests of Sharpie's specimens were carried out using the IO-5003 installation. The steels strength and plasticity characteristics were determined by the results of tension tests of the smooth specimens with a cross section 4×5 mm using UME-10T installation. Specimens were oriented across of the rolling direction of the elements with angular profile and tested in the air and in an artificial acid rain environment (SCC test) at the deformation rate of $3{\cdot}10^{-3}$ and $1.7{\cdot}10^{-7}$ s^{-1} correspondently. The composition of this environment (in g/l: Na_2SO_4 (3.2); HNO_3 (1.58); $NaNO_3$ (2.12); NaCl (8.48), with the addition of H2SO$_4$ to achieve pH 4.4 in the resulting solution) was chosen basing on the analysis of the rainwater composition in the industrial regions. Metallographic examinations of structure been carried out by an EVO-40XVP scanning electron microscope. The reagent consisting of 3% nitric acid solution was used for etching of the metal structure. The fractographic investigations of the fracture surfaces of tested specimens were also carried out by an EVO-40XVP scanning electron microscope.

4.3 Results and Discussion

4.3.1 Typical Corrosion Damages of the Different Elements of Shukhov's Tower in Mykolayiv

The corrosion damages on the surface of the tower's elements (especially, near zones of their connecting by rivets) caused by their long term exploitation were analyzed. In the postwar period a critically damaged elements were replaced by new elements with angle profile (in further named as repair elements) using welding technology for them connecting. As a result, an additional factor of the influence on the structural integrity of the tower was appeared associating with workability of the weld joints.

The longitudinal corrosion delaminiations were observed throughout the entire thickness of the connected steel elements. These were typical corrosion damages observed at the ends of all elements of the tower (Fig. 4.5a). In most cases the deformation of all cross section of both elements was observed in the places of their connecting. Significantly larger volume of corrosion products relative to ferrite matrix had caused such deformation effect. Observed delaminations are also connected with an existence in the steels structure of the chains with non-metallic inclusions oriented along the rolling direction of the corners. Old steel with a large number of non-metallic inclusions is more susceptible to corrosion degradation due to the greater electrochemical heterogeneity of its structure. The corrosive damages to all elements of the tower were caused by the rainwater. At the same time, the water remained for a longer time on the horizontal shelves of the angular rolled metal and in the gaps between the joined elements adjacent not tight enough to each other. As a result the favorable conditions for both general and local corrosion on the horizontally oriented surfaces of the elements, as well as for SCC of the weld joins and anchor bolts for fixing the tower on the concrete foundation were created (Fig. 4.5b). Penetration and long-term holding of water in the stress concentration zones existing in the weld joints and in threaded joints of the anchor bolts and nuts could promote their destruction

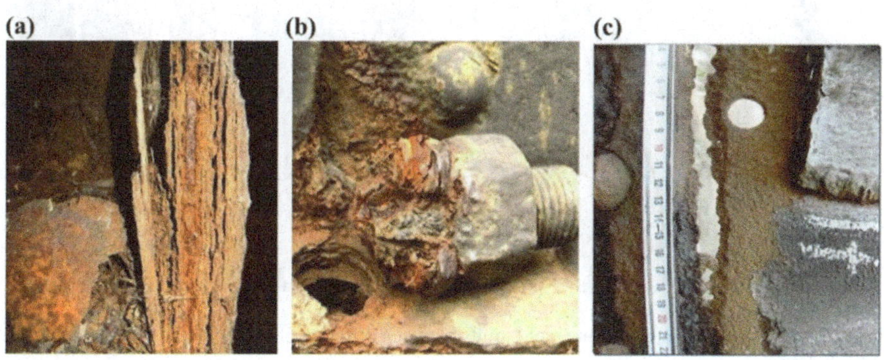

Fig. 4.5 General view of the typical corrosion damages of elements Shukhov's tower in Mykolayiv

by realization of the mechanism of SCC. The deep and quite often through cross section corrosion damages of the elements of lower horizontal rim of the Shukhov's tower were revealed (Fig. 4.5c). Such extensive damages were made possible due to a long-term delayed of the acid containing precipitations (especially at the junctions of two elements).

Particular attention during inspecting of the tower was given to the inspection of the state of welded joints made during repair and restoration work in the post-war period. All investigated welds used to replace rivet joints between the support bars with angular profile and horizontal rims of the tower were completely (Fig. 4.6a) or partially damaged (Fig. 4.6b, c). The intermittent morphology of macrocrack along the weld line of one of the elements with a partial separation from the rim enable to assume that these cracks along the repair weld occurred as a result of SCC (Fig. 4.6c). It was believed that significant residual stresses, most likely due to forced pressing of welded elements using additional clamping devices. The favorable conditions for the realization of the SCC could be created at the periodic contact of welded elements with acid-containing sediments. It was believed that not only the imperfection of welding technology, but, first of all, SCC is the true cause of damage to most welded joints. At the same time, authentic joints of elements using rivets continue to ensure the stability of the tower as a whole until now.

It was also noted the presence of authentic elements of the underground communications of the tower in Mykolayiv (Fig. 4.7a). It is rather an exception from the rules for still extant of the Shukhov towers. Although these elements are significantly damaged by corrosion and some from these components had end-to-end damages, but as a whole the underground communications give an idea of the constructural decisions and joins between the elements used at that time. Bolts and nuts used

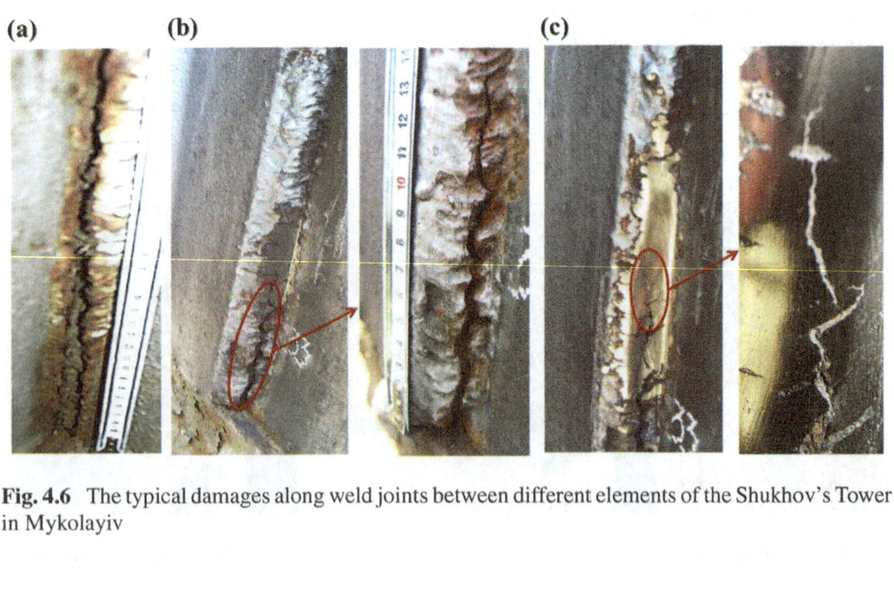

Fig. 4.6 The typical damages along weld joints between different elements of the Shukhov's Tower in Mykolayiv

Fig. 4.7 Typical corrosion damages of the elements of underground communications to ensure the supply of water to the collector at the top of the tower

for flange connections are almost completely destroyed by corrosion and the nuts practically break up into fragments (Fig. 4.7b).

It was detected that a piece of the pipe used for the filling of the water collector at the top of the tower by water was bulged out. It may indicate on the possible freezing of water inside the pipe. A larger volume of ice than of the water caused the expansion of the pipe walls from the inside with subsequent its destruction (Fig. 4.8).

4.3.2 The Composition and Hardness of the Steels of Elements from Shukhov's Tower

The chemical compositions of an old steel (1906) and the repair one (1944) and of the correspon^{2}ding them modern medium carbon steels are presented in Table 4.1. The criterion for choosing prototypes for both exploited steels was the approximate correspondence of the content of carbon in them.

Old and repair steels from different elements of the tower differed in distribution of the hardness values (Fig. 4.9). The hardness of the repair steel from different

Fig. 4.8 The damaged pipe used for the collector filling at the top of the tower by water and located above the ground

Table 4.1 The chemical composition of steels from the Shukhov's tower elements and of their modern prototypes

Content of elements, wt. %						
steel	C	Mn	Si	Cu	S	P
Old	0,28	0,48	0,006	0,013	0,043	0,019
Repair	0,34	0,64	0,270	0,104	0,066	0,038
St3	0,22	0,07	0,06	0,026	0,035	0,02
Steel 35	0,38	0,26	0,64	0,023	0,028	0.05

elements of tower was more stable, and its values ranged from 112 to 123 HB (the average value of the 100 measurements was 118 HB). The hardness of old steel was much lower. When hardness of the various elements ranged from 64 to 92 HB, the average value was 83 HB. First, the lower hardness of the old steel is caused by the slightly lower carbon content in it. In addition, it can also be caused by the scattered damages inside the metal which formed during over more than 100 years of service of old steel. The similar change in hardness due to the scattered damages inside the steels and welded joints after their long-term operation in other structural elements and service conditions was earlier detected [25, 26].

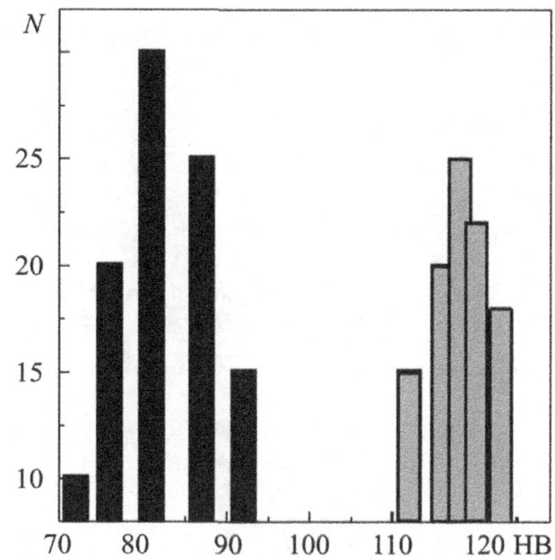

Fig. 4.9 The distributions of the hardness HB values measured on the different elements of the tower made from the old steel (black columns) and the repair one (grey columns)

4.3.3 Microstructure Features of the Investigated Steels

Basing on the metallographic analysis was shown that ferrite-perlite structure was typical to both steels of the elements of the Shukhov Tower (Fig. 4.10). The fraction of the perlite is higher in the repair steel compared to the old one due to the higher carbon content in them. The texture in the rolling direction on the longitudinal sections of the elements from the repair steel had manifested more clearly than in the old one. In last case the rows of perlite grains were oriented in the direction of rolling only as an exception to the rule.

Usually, in the old steel, the perlite grains had an irregular elongated shape and extended along the boundaries of the ferrite grains, both in the longitudinal and in the transverse section of the elements. In repair steel, the texture in the longitudinal section of the elements is more pronounced, the shape irregularity of the perlite grains is somewhat smaller (especially in the transverse section of the elements), their dimensions are larger (regardless of section orientation in analyzed element). At the microstructural level, no differences were observed between the characteristic dimensions (length and thickness) of cementite plates inside perlite grains of both steels. Their dimensions were almost the same in both steels and did not depend on the orientation of the analyzed cross-sections of the elements for metallographic studies. In general, the thickness of the cementite lamellas varied in the range from 0.2 to 0.6 μm, and their length was from 0.5 to 10.0 μm (Fig. 4.11a, b). In the repair steel longer cementite lamellas was prevailed within the perlite grains (Fig. 4.11c). While in the old steel the opposite trend was observed. It was explained by the essential differences in the sizes of the perlite grains in these steels and their elongated shape in the old steel.

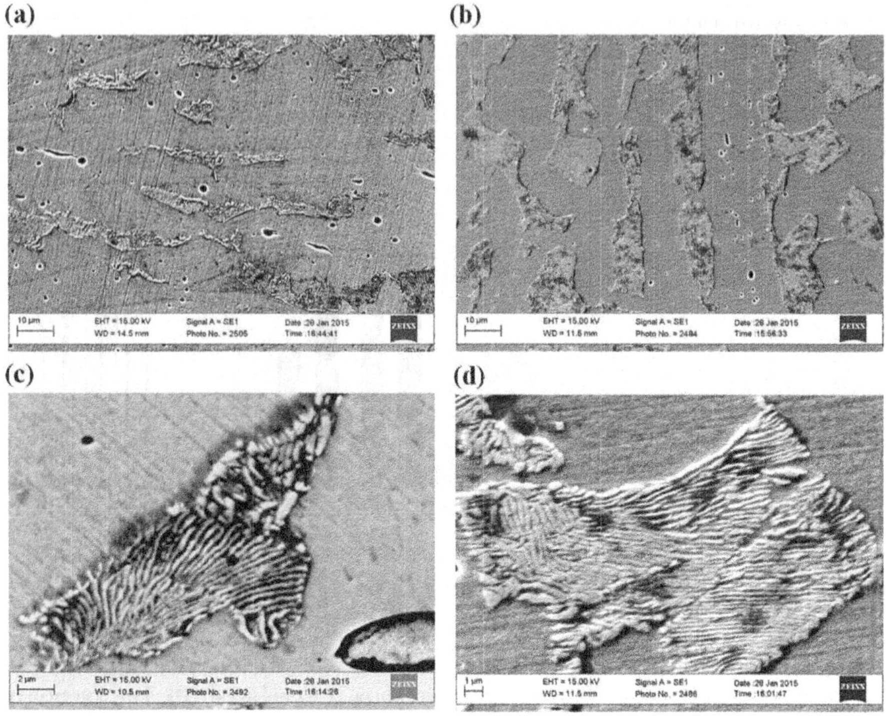

Fig. 4.10 The microstructures of the **a, c** old steel and **b, d** of the repair one from the elements of the Shukhov's Tower

The traces of nonmetallic inclusions oriented in the rolling direction of elements were observed in the structures of both exploited steels. These tracks were mainly observed on the background of the ferrite grains. Moreover the sizes and numbers of the nonmetallic inclusions in the old steel were larger than in the structure of the repair steel. In the most cases, these inclusions were practically not found inside the pores on metallographic sections from both exploited steels. It was concluded that their connections to the ferrite matrix were weak. Such traces left by inclusions were found in both steels. This structural feature was considered as a sign of the scattered damages inside both steels formed during their exploitation. The nonmetallic inclusions are usually remained in the places of their locations in structures of the non-operated steels (even at intensive polishing of the metallographic specimens). However, at the structural level the operational degradation of the old steel was more visible than of the repair one. This is the result of joint influence of a longer service life of old steel and more numbers of non-metallic inclusions in its structure. As a consequence, at the maintaining of the integrity of the old steel at the macro level the extensive network of damages had formed inside it at the micro level. The mechanical characteristics of the porous metal will certainly differ from the corresponding characteristics of the defect-free metal. It will be effect on the

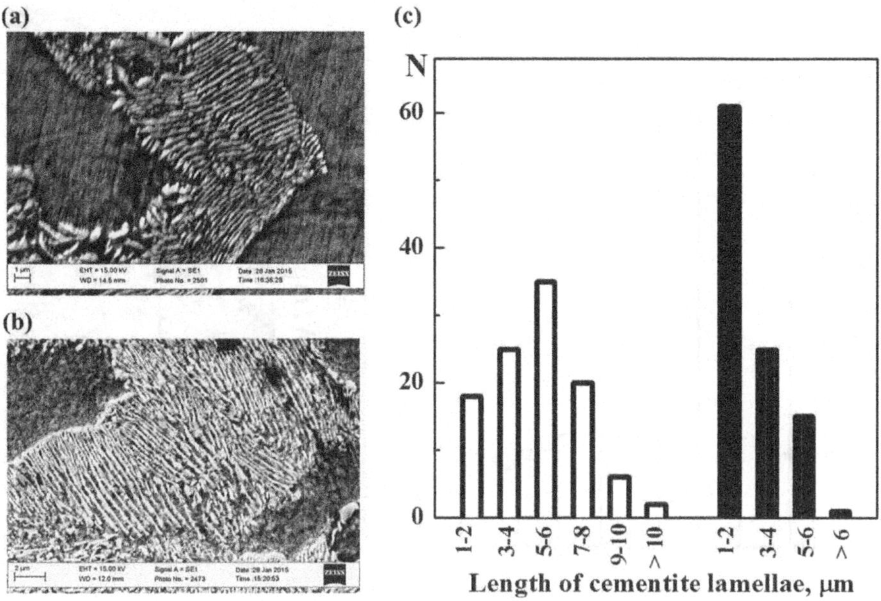

Fig. 4.11 The structure of an **a** old and **b** repair steels from elements of the Shukhov's tower and **c** the distribution of lengths of cementite lamellaes inside the perlite grains in the repair steel (*white*) and in the old one (*black collumns*)

operability of the exploited elements of the tower, especially under the influence of an acid rain, which can't be avoided in the industrialized regions.

4.3.4 Brittle Fracture Resistance of the Exploited Steels

The results of the assessment of resistance to brittle fracture of both the exploited steel (old and repair one) and their modern prototype (steel St3) are presented in Table 4.2.

Obviously, due to the long-term operation of steels in the elements of the Shukhov's Tower, the ability of the old steel to resist the influence of the power impact was the smallest. The reduction in toughness of the steel used in comparison with non-operating steel St3 (used as a modern prototype for both exploited steels) reached more than 85% and about 70% for old and repair steels, respectively. Thus, the impact toughness values of the exploited steels unambiguously reduced as the

Table 4.2 Impact toughness *KCV* of the investigated steels

Steel	Old	Repair	St3
KCV, MJ/m^2	0.15	0.3	0.98

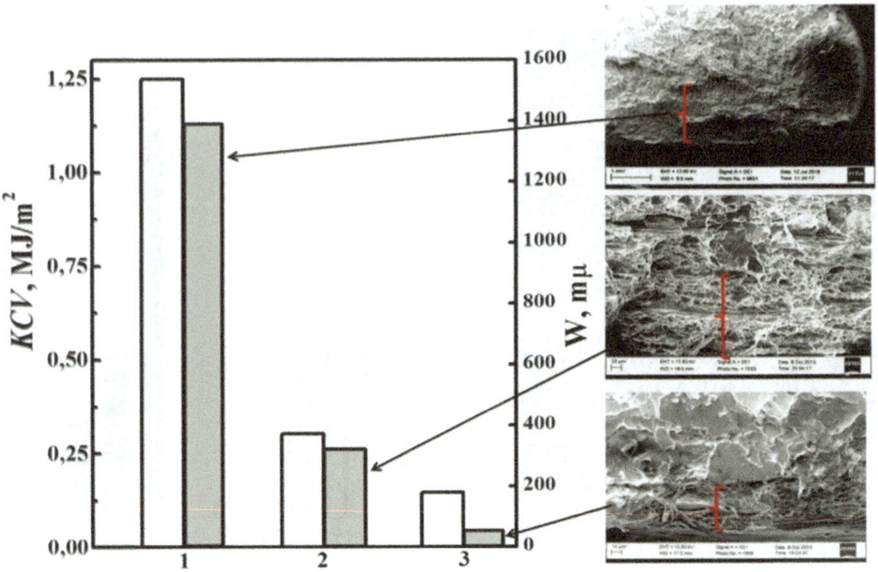

Fig. 4.12 The values of the impact toughness KCV (*white columns*) and of the average width W of the ductile failure zones in the vicinity of the stress concentrator (*gray columns*) for the steel St3 in the virgin state (1) and also for the repair steel (2) and for the old one (3) after their long-term exploitation in the structural elements of the Shukhov's Tower

duration of their operation in the elements of the tower increases. The KCV values of all investigated steels correlates with the average width (through the specimen thickness) of the ductile failure zones in the vicinity of the stress concentrators on the fracture surfaces of the corresponding specimens (Fig. 4.12).

It was believed, that a significant reduction in the resistance to brittle fracture of exploited steels compared to modern steel was associated not only with the inevitable differences in production technologies, but also with the degradation of the old and repair steels during their long-term operation in the elements of the Shukhov's Tower. Certainly such a significant decrease in the impact toughness of the exploited steels must be taken into account (especially for justification of the workability of the tower as a whole).

4.3.5 Fractography Features of the Impact Specimens

In general, the fracture surfaces of impact specimens of the old and repair steels on the macro level were flat and without stretching lips. Such distinctive elements of the fracture surfaces are the macro signs of the brittle fracture beginning in the central part through the thickness of the specimens. Traces in the form of macroscopic crests of the tear off were clearly distinguished on the fracture surfaces in the form of fan-like pattern starting from the points of brittle fracture initiation (Fig. 4.13a). A feature of the fracture of the repair steel was a wide zone (at maximum width of 0.5 mm) with characteristic signs of the separation along the interfaces of the structural components (Fig. 4.13b). Considering the high (up to 0.6%) content of manganese in the repair steel, was assumed that these traces in the form of perforated bands, located along the rolling direction remained on the fracture surface due to separation of the manganese sulphides from the ferrite matrix.

The visible elongated traces formed due to delamination along the interfaces between non-metallic inclusions and the matrix (Fig. 4.14b) on the background of the typical ductile relief from the dimples appearing due to the shearing and tearing processes in the pre-fracture zones were revealed on the fracture surfaces of the specimens of both steels within ductile zone of failure (Fig. 4.14a, b). Inclusions that triggered the appearance of both types of dimples on the fracture surfaces were mostly absent on their bottom. This feature was observed at the fracture surfaces of both steels, but in the case of the old steel this was manifested with practically no exceptions. It is quite possible that a partial destruction of cohesion between non-metallic inclusions and a ferrite matrix during the long-term operation of both steels in the elements of the Shukhov's Tower was finally completed during impact tests of specimens from them. As a result, only the traces left by these inclusions were observed at the fracture surfases of the impact specimens of both steels. We also noted a higher relief of the fracture surface of the repair steel at the initial stage of ductile fracture compared with observed on the old steel. This indicates on the higher energy-intensive destruction of repair steel and is consistent with the results of impact tests.

Moreover the separation traces of small cementite particles within perlite grains were also observed on the fracture surfaces of both steels within ductile zones of failure (Fig. 4.14b, c). In the old steel, even the fine cementite particles were not observed between the stretch crests along their interfaces with ferrite lammelas, whereas in the repair steel most of them still retained the connection to the matrix.

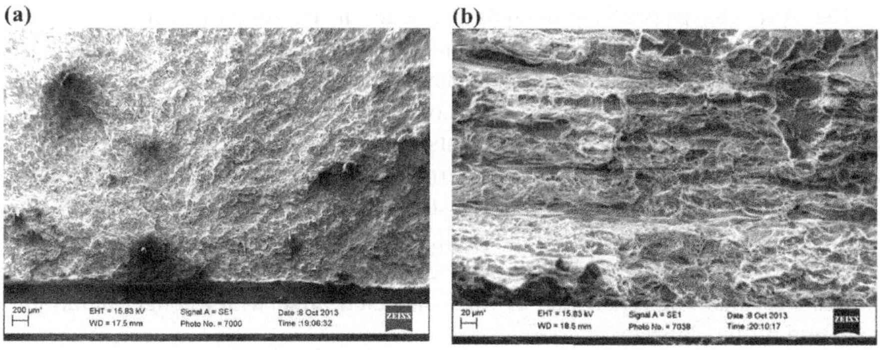

Fig. 4.13 Macrofractograms of the old steel (**a**) and the repair one (**b**) from the Shukhov's Tower elements observed on the fracture surfaces of the impact specimens in the vicinity of their stress concentrators

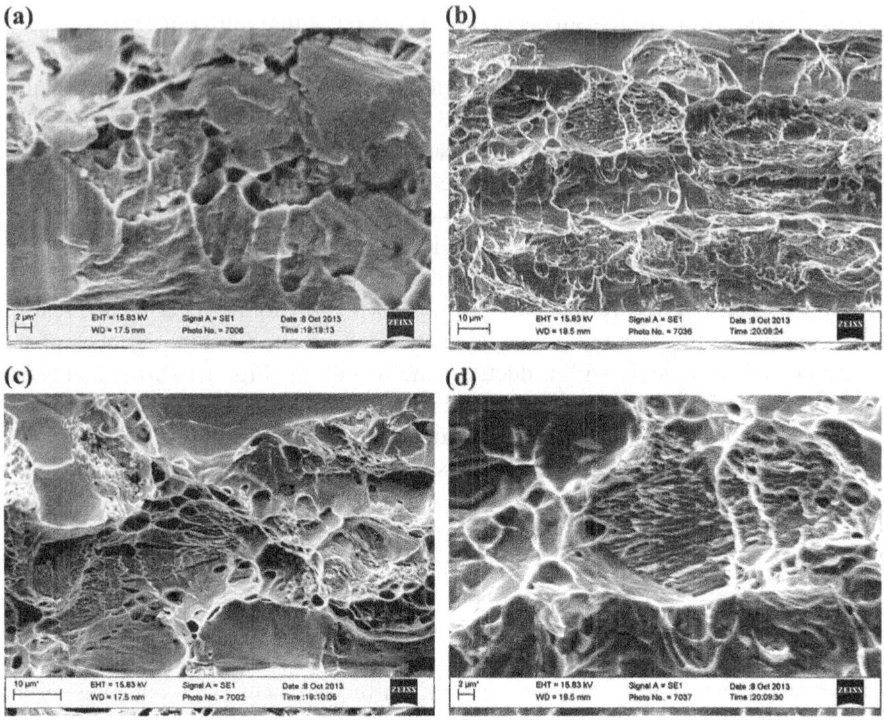

Fig. 4.14 Microfractograms of the old steel (**a**, **c**) and the repair one (**b**, **d**) from the Shukhov's Tower elements on the initial ductile stage of fracture of the impact specimens

This gives reason to say that damages of the old steel during long-term exploita-tion occurred not only along the nonmetallic inclusions with the matrix, but also inside the interfacial interfaces inside the perlite grains. Whereas the damages process inside perlite grains was not yet completed in the repair steel.

The most unexpected fractography elements in the zone of ductile fracture of impact specimens of the both steels became the classical brittle transgranular facets with the relief by crests formed the typical river patterns which clearly distinguished on the background of the dimpled relief of the fracture surfaces (Fig. 4.15a, b). It was assumed that these cleavage elements were initiated by damages in the metal structure that occurred during the long-term operation of the steels. In modern steel St3, chosen as a prototype for exploited steels, such cleavage elements in the zone of ductile fracture were not observed at all. The cleavage regions on the background of the ductile relief covered by dimples were more often observed at the fracture of the old steel than the repair one. It was connected, first, with a longer duration of operation of the old steel.

The ratio of the areas with the cleavage and ductile fracture in the zones of the initial stage of ductile failure of the impact specimens in the vicinity of their stress concentrators were more than twice as much in the case of old steel as in the repair one

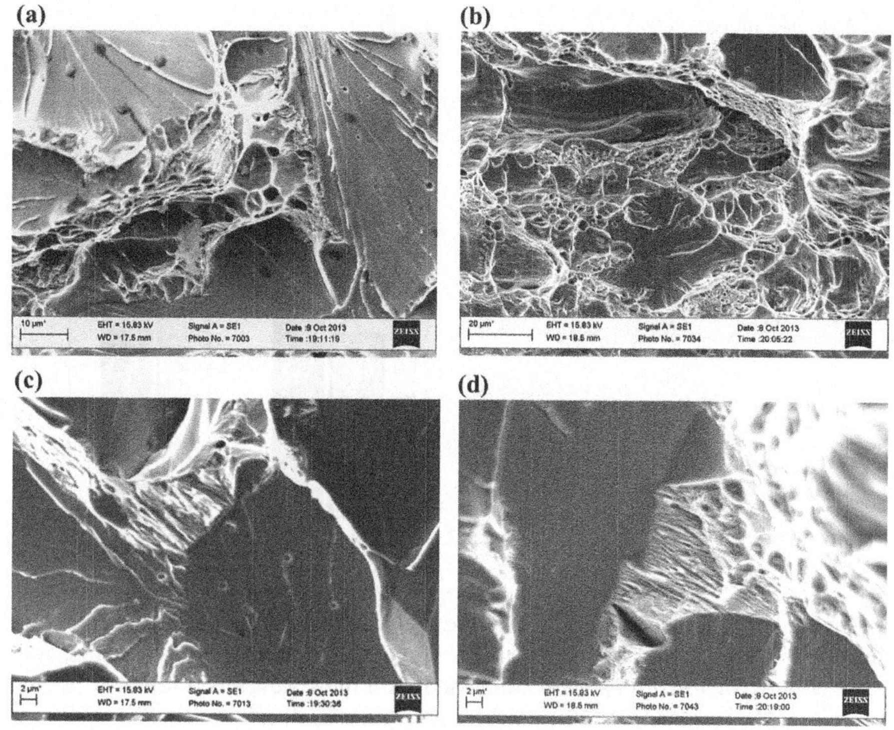

Fig. 4.15 Microfractograms of the old steel (**a**, **c**) and the repair one (**b**, **d**) from the Shukhov's Tower elements on the initial (**a**, **b**) and final (**c**, **d**) stage of the impact specimens fracture

(Fig. 4.16). These fractography valuations are in a good agreement with a double difference in the impact toughness between these steels. After all, the more such cleavage elements on the fracture surface (in our case, this is typical of old steel), the lower the impact toughness of steel.

At the final stage of specimens failure of both steels (of the old and the repair one), a transgranular brittle fracture was prevailed (Fig. 4.15c, d). However, at this stage of the destruction of specimens the some features of the relief were also noted. The boundaries between the structural components within perlitic grains (cementite and ferrite) were clearly distinguished at the fracture surfaces of both steels (Fig. 4.15c, d). It was their common feature in the zone of spontaneous destruction. At the same time, such a clear separation of components of perlite was practically not observed at the failure of the impact specimens from the modern steel in the initial state. It was assumed the connection between the structural components of the perlite had weakened during long-term operation of both steels in the elements of the Shukhov's Tower. As a result, the decohesion effect (at least partial) had observed at the fracture surfaces of the impact specimens from exploited steels in the form of clearly delineated boundaries between the structural components of the perlite.

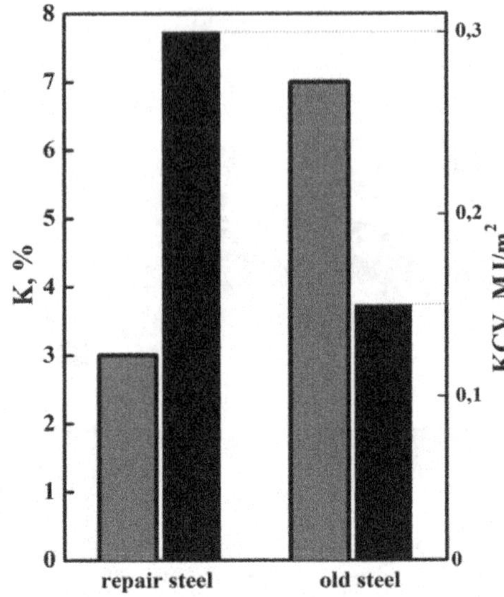

Fig. 4.16 The ratio (K) of areas of cleavage and ductile fracture in the zone of the initial stage of ductile failure of the impact specimens in the vicinity of their stress concentrator (*gray columns*) and corresponding them the impact toughness values KCV (*black columns*) estimated for the repair steel and for the old one

The higher energy consumption for the specimen's destruction of the repair steel (in comparison with the old steel) was also manifested in the stage of brittle transgranular fracture. Such features were connected with more developed relief of fracture surfaces at the transition from grain to grain (on microscale level) and with large number of the elongated sites on fracture surfaces with significant displacement relative to each other (on the mesoscale level). As a consequence, the connection of the spatially displaced sites on the fracture surface was occurred by significant plastic deformation of the bridges between them. Therefore, at the interfaces between sites with purely brittle fractures, were often observed the ridges covered by an equiaxial dimples formed during deformation of these bridges during coalescence of the adjacent cleavage areas. On fracture surface of the old steel the number and the length of similar crests covered by the ductile dimples were significantly smaller than in the repair steel. Therefore the ability of the old steel to resist by the dynamic overloading is significantly less than of the repair one.

Thus, the comparative fractographic analysis of the features of the destruction mechanisms of impact specimens of the old and repair steels, had confirmed that the resistance to brittle fracture of old steel is substantially less than that of repair steel. Consequently, the fractographic studies also confirmed that ability of the old steel to relaxing the stresses arisen in the elements of the Shukhov's Tower at the impact loadings is stronger exhausted than of the repair steel. It is generally known that the lower the impact toughness levels of steel, the stronger its tendency to the embrittlement effect of the corrosive environment. Therefore, in order to substantiate the possibility of further operation of the tower, it is important to assess the ability of both steels (with their unusually low resistance to brittle fracture) to resist to the combined action of stresses and corrosive environment.

4.3.6 Mechanical Properties of Steels at Tensile Test

The analyze of strength characteristics of both steels after tension test in the air were shown that the σ_{YS} and σ_{UTS} values of the repair steel were higher compare to the old one (Table 4.3). It was also identified an atypical change in the plasticity characteristics. Despite the practical invariance of the reduction area values of both steels tested in air, the elongation value of the old steel slightly increased in comparison with the corresponding value of the repair steel. The revealed tendency of the plasticity characteristics changes could be associated with the smaller number of the pores revealed in repair steel compare to the old one. The opening of the defects arisen inside of the steels during their long-term operation was additionally increased at the stage of uniform elongation of specimens and, as a result, the elongations of specimens from degraded steels were additionally increased. Usually the both of the plasticity characteristics of steels are changed in the same direction (together or decreasing, or increasing) under the influence of any other factors (such as heat treatment or alloying). The trend of their opposite change is only inherent to degraded steels. The similar trend of the atypical change in the plasticity characteristics was previously found on the heat resistance steels of the main steam pipelines of the thermal power plant after a different number of process shut-down during their exploitation [15, 27], and on the exploited weld metal [25], and on the steels exploited on the main pipelines for gas and oil transportation [28, 29].

During long operation, the elements of the tower were subjected to joint influence of the seasonal precipitation (rain, snow) and different stresses (in particular, of the static stresses, due to the weight of the reservoir with water at the top of the tower, of the dynamic stresses due to wind gusts and of the cyclic stresses due to temperature changes during the days and seasons). The surface corrosive damages and delaminations of metal along the texture in the rolling direction at the end faces of all tower elements had been revealed in the places with favorable conditions for long-term moisture holding (Fig. 4.5a). Moreover, the high electrochemical heterogeneity on the microstructural level is typical for both steels. It is caused by the large number of non-metallic inclusions in them (Fig. 4.10a). These inclusions can be efficient cathodes during electrochemical corrosion process at the interaction of a metal with an aqueous environment. The localization of the cathode process on nonmetallic inclusions is accompanied by the formation of hydrogen (as one of the reaction products) and promotes hydrogenation of the metal in the vicinity of inclusions [30–33]. The corrosion resistance of steel is significantly reduced under the influence

Table 4.3 Mechanical characteristics of investigated steels at tension test in air

Steel		σ_{UTS}, MPa	σ_{YS}, MPa	Elongation, %	RA, %
Old	Exploited	482	310	26,3	41,0
Repair		637	424	22,4	41,7
St 3	Initial state	497	310	28,3	61,5
Steel 35		656	430	23,4	54,5

of hydrogen [34]. In addition, hydrogen absorbed by the metal can localized in the vicinity of the nonmetallic inclusions and to facilitate their separation from the matrix with forming the pores along their interfaces. Inside these pores, hydrogen transforms into the molecular state and its pressure increases with further accumulation in the pores. Local stresses into the formed cavities (due to hydrogen pressure inside them) and stress concentration in the vicinity of nonmetallic inclusions (especially with disc like shape) can even cause hydrogen induced cracking (HIC) of the steel. Low values of the impact toughness of degraded steels after their long-term operation usually create the favorable preconditions for the successful implementation of this failure mechanism. The results of the impact tests of both steels exploited in the elements of Shukhov's Tower also testifies the favorable conditions for their hydrogen embrittlement (Table 4.2) [34]. It is especially visible in the case of old steel with a critically low impact toughness value and with much more number of the non-metallic inclusions in it.

With accounting of the possibility of hydrogenation of tower elements due to the electrochemical heterogeneity at the micro-structural level the determination of the mechanical characteristics of both steels (of the old and repair one) after the SCC with SSRT of the smooth specimens in the artificial acid rain environment with pH4.4 is also important. Thus, the influence of the atmospheric environment on the loaded of the tower elements was simulated.

The relative change in the corresponding mechanical characteristics of the old steel to respect of repair one was estimated using of the λ ratios (Fig. 4.17). The λ values for both steels tested were obtained using the result their test in air and in the artificial acid rain environment. These ratios characterize the effect of degradation of the corresponding mechanical characteristics of old steel in relation to the repair one at testing of specimens on air and after the SCC with SSRT in the corrosion environment. It was noted that the strength and one of the characteristics of plasticity

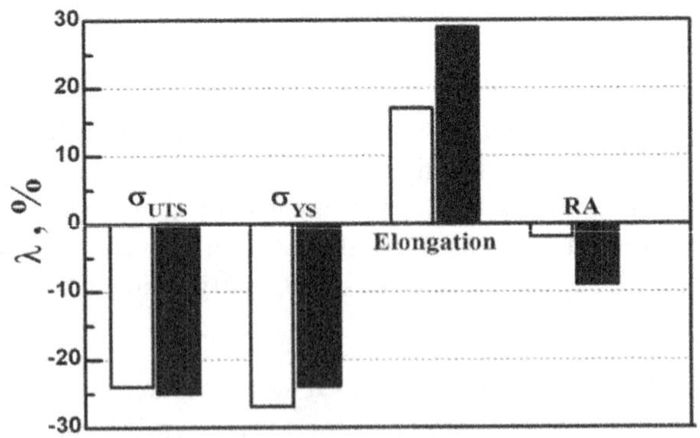

Fig. 4.17 Effect of the degradation degree of the old steel (due to a longer service life in the elements of the Shukhov's Tower) regarding to the degradation of the repair steel evaluated by using the λ ratios which characterized the relative change of the corresponding characteristics (of strength and ductility) of both steels, determined in air (*white*) and in the environment (*black columns*)

(namely the reduction of area) of the old steel were lower than the repair one (both in the air and in the environment). It may be a consequence not only of a more advanced technology for manufacturing repair steel that provided its better characteristics in the virgin state, but also by the significant degradation of the old steel during longer service life. Moreover an atypical change in the elongation values was noted. The elongation values for old steel are higher than for the repair one (during tests both in air and in aggressive environments). As already mentioned above, this effect can be associated with appearance of structurally caused of the randomly scattered damages in the steels formed during their long-term operation. With an active load of the specimens, these damages inside their working areas had further opened and, thus, increased of the specimen's elongation. The area reduction values of the old steel were predictably lower than of repair one, but its elongation values abnormally increased in both test environments. The opposite change of the plasticity characteristics caused by the degradation of the steels is much more obvious on the specimens tested in the hydrogenated environment. Thereby, it was also shown the increasing of the sensitivity of the plasticity characteristics to the operational degradation of steels by using the embrittlement effect in the artificial acid rain environment.

It is known that hydrogen could accelerate the degradation of steels [35–37]. Since the tower exploitation was accompanied by the combined action of mechanical loads and of the hydrogen absorbed by the metal during electrochemical interaction with the corrosive environment, can be expected the similar effect of accelerated degradation of steels from tower elements under the influence of hydrogen. On the one hand, it can intensify the diffusion redistribution of elements in the metal [38–40]. On the other hand, the hydrogen can facilitate the formation and development of damages inside the metal [25]. Analysis of the known manifestations of exploited degradation of structural steels indicated that the resistance to brittle fracture of exploited steels is especially strong reduced. Consequently, degraded materials become particularly sensitive to corrosion-mechanical influences. The risk of uncontrolled fracture of the operated materials is increased at the influence of the corrosive environments. Their resistance for the corrosion-mechanical failure is evidently decreased. The risk of degraded metal destruction can increase in corrosive environments for two reasons. Firstly, due to ability of the corrosive environment intensify the hydrogen degradation of metal during its long-term operation. And secondly, due to its ability facilitate the fracture process by realization of the SCC or corrosion-fatigue crack growth [41, 42].

To separate the corrosive environment effect on the mechanical characteristics of each of the studied steels, their characteristics of strength and plastic-ity obtained after the tensile tests in the artificial acid rain environment and on air were compared (Fig. 4.18). It was achieved using the λ ratios, which determined as the difference between the corresponding characteristics for analyzed steel in the both environments relatively to its value determined on the air. It was revealed that the sensitivity of the strength and plasticity characteristics (by using of the λ ratio) of the exploited old and repair steels to impact of the acid rain is higher than of the corresponding steels in the initial state which used as the prototypes. Strength characteristics and area reduction of specimens were shown low sensitivity to environmental effects.

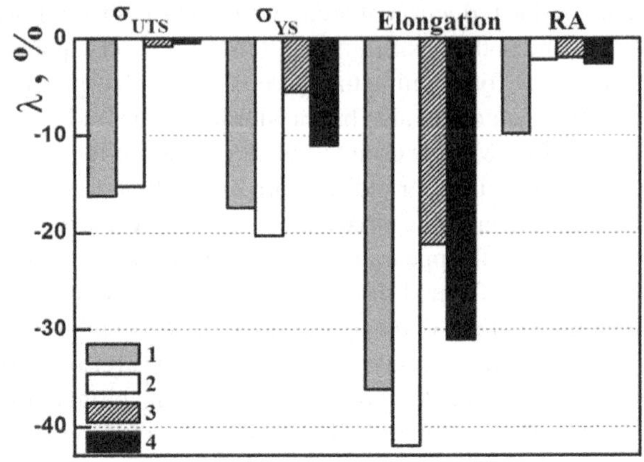

Fig. 4.18 The corrosion environment effect on the strength (σ_{UTS} and σ_{YS}) and plasticity (elongation, RA) characteristics estimated using the λ ratios showing the change in the corresponding characteristics of the degraded steels (old (1) and repair (2) ones) and of their non-exploited prototypes (the St3 (3) and St35 (4) steels) obtained in the artificial acid rain environment with respect to those obtained in air

In particular, under the influence of acid rain the maximum reduction in the tensile strength value of the exploited steels was 17%, and of the yield strength—20%. While under the influence of this environment, the values of σ_{UTS} of non-exploited steels remained practically unchanged, and of σ_{YS}—decreased by the 11%. A more significant decrease in the strength characteristics of the operated steels compared with the non-exploited ones was associated with randomly scattered damages inside the metal which formed during its long-term operation. It is clear that additional loss of steels strength under the influence of acid rain must be taken into account during calculations of such constructor's integrity.

The change of the reduction area values of specimens from the repair steel and from both non-exploited steels did not exceed the experimental error, and only for the old steel this value several decreased (by 10%). The elongation values of the samples from all considered steels were sensitive to the action of acid rain. Acid rain had affected not only on the exploited steels, but also on their prototypes in the original state. In particular, the influence of the environment on the specimen elongations (with using of the λ ratio) were of 21% and 31% for the modern steels with carbon contents of 0.2% and 0.35%, respectively. Due to the degradation of the steels of the elements of the Shukhov's Tower after their prolonged operation, the relative elongation of the old steel decreased under the influence an acid rain on the 36%, and of the repair steel—on the 42%. Surprisingly, but the sensitivity to the action of the environment of the old steel for the value of specimen elongation was somewhat lower than of the repair one. Perhaps this feature is associated with lower carbon content in old steel compared to repair one. After all, the higher is the carbon content in steel, the higher is its sensitivity to the environment effect. And this was clearly confirmed by the results obtained on non-exploited steels.

Thus, this analysis clearly demonstrated that the mechanical characteristics of degraded steels are more sensitive to acid rain influence than of the steels in the initial state. It is clear that this must be taken into account during calculation of the Shukhov's Tower on it strength in general.

The characteristics of strength and ductility of the old and repair steels were compared to corresponding properties of their modern prototypes (the St3 and 35 steels). To simplify the analysis of the degradation effect, the results were presented using the λ ratios characterizing the change in the characteristics of the exploited steels with respect to the corresponding characteristics of the prototype steels in the initial state (Fig. 4.19). The obtained strength (σ_{UTS} and σ_{YS}) and ductility (elongation, RA) characteristics indicate a more intensity degradation of the old steel, regardless of the test environment (in air or in an acid rain). The negative effect of the degradation of both steels (the old and repair ones) on all analyzed characteristics was significantly increased during the testing of specimens in the artificial acid rain environment, compared to observed in air. And in all respects, the degradation of the old steel was much stronger than of the repair steel. Therefore, to distinguish the effect of steel degradation, as the prototype for comparison the non-exploited steel of the appropriate composition (at least with account of the carbon content) is necessary used.

It was also noted, an unusually high sensitivity of the RA characteristics to the steel degradation. It is a distinctive feature of the steels exploited in the tower elements in compare with the modern steels in the initial state with the corresponding carbon content. First of all, it concerns of the old steel. In this case the degradation effect by the characteristic of RA was almost 40%, when specimens tested in a corrosive environment.

Typically, the sensitivity to the degradation of the heat resistance steels of such characteristics as the relative elongation of specimens is significantly higher than of the area reduction values [15]. However, in the case of tests of ordinary carbon steels, which were exploited for a long time under the influence of climatic temperatures

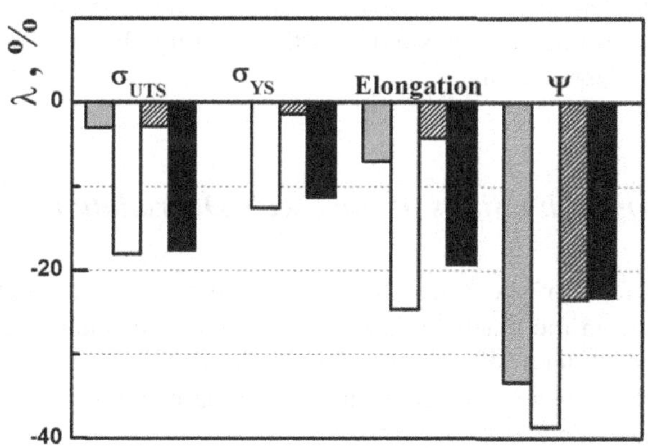

Fig. 4.19 The degradation effect of the strength (σ_{UTS} та σ_{YS}) and plasticity (elongation, RA) characteristics due to long-term operation of old (marked by the grey and white columns) and repair (by shaded and black ones) steels in the elements of the Shukhov's Tower regarding to the corresponding characteristics of the prototypes for these steels in virgin state evaluated by using the λ ratios which characterized their relative change at determining in air (*grey and hatched columns*) and in the artificial acid rain environment (*white and black columns*)

and environments, a stronger change in their reduction of area values compared with the elongations ones due to the steels degradation was also noted [43, 44]. The obvious difference in the tendencies of change in the plasticity characteristics due to the operational degradation of the low-alloyed heat resistant steels and of the carbon ones was explained by the different nature of the processes of damaging during high- and low-temperature exploitation. In the first case, the degradation is due to favourable temperature and force conditions for diffusional redistribution of elements with the formation of large carbides and along the grain boundaries pores around them due to creep. In the second case, when the diffusion redistribution of most of the elements is limited (by low service temperature), the degradation was associated with local electrochemical reactions with the formation of hydrogen due to structural heterogeneity of steels. Due to these reactions inside the metal the rather large defects (compared to the pores around carbides) in the form of delaminations between non-metallic inclusions and the ferritic matrix (or along the interfaces between the ferritic and perlitic grains) were formed. Thus, a large number of relatively small damages in the form of pores (scattered at the micro level inside the metal) formed during high-temperature exploitation of the steels, additional open during tensile testing of the specimens from the exploited metal and affect on their overall elongation, but have a little effect on the localization of the specimens deformation and on their values of the reduction of area. As a result, the weak sensitivity of the reduction of area values of the heat resistant steels to degradation was observed. The degradation of steels at low-temperature exploitation is manifested by more extensive damages (commensurable with the size of structural components in the steels), corresponding to the scattered damages on the mesodimensional level. The open of these damages will undoubtedly contribute to the localization of the deformation during tensile tests of the specimens and, consequently, to a noticeable decrease in their values of the reduction of area. Therefore, the level the scattered damages inside the metal (micro- or meso-) predetermines the sensitivity of the ductility characteristics of steels to their operational degradation.

4.3.7 Fractography Signs of the Steels Degradation

The fracture surfaces of specimens from the old and the repair steels after tension testing on an air, on the macro dimensional level had the characteristic geometry of the cups type. After testing both steels in an artificial acid rain environment, the fracture surfaces were oriented by normal to the macro direction of specimens loading (without evident conical parts).

The typical ductile fracture predominated in the central parts of the specimen surfaces of both steels tested in air, where destruction of smooth specimens usually begins. During the long operation of the steels the favourable conditions are created for nucleation, growth and coalescence of the voids around non-metallic inclusions due to the reduction in their cohesion with matrix. In consequence of the coalescence of micropores by the deforming the bridges from the intact metal between them until to their destruction the dimple like relief were formed (Fig. 4.20a, b). The most dominant size of the inclusions at the fracture surface of old steel was 3...4 μm, and of repair one −1...2 μm (Fig. 4.20c). As a result the dimples of the significantly larger sizes were observed on the fracture surfaces of the old steel, compared with the repair one. It was noted that most often non-metallic inclusions kept on their own locations at the bottom of the dimples. This indicates in the preservation (at least partially) of their cohesion with the matrix after so long exploitation.

In both cases the dimples were also formed around the cementite lamellas within the pearlite grains and the ferrite bridges between them stretched and ruptured with formation of the small separation ridges (Fig. 4.21). The shape of the small dimples changed depending on the orientation of the cementite lamellas regarding to the fracture surface. The elongated dimples inside the perlite grains were formed when cementite lamellas were located at an angle to the fracture surfaces of specimens (Fig. 4.21a). The fine equiaxial dimples were observed inside perlite grains when cementite lamellas were practically oriented perpendicular to the fracture surfaces of specimens (Fig. 4.21b). In any case, their dimensions were commensurate with thickness or cross section of the cementite plates. In the vicinity of the side surfaces of the specimens after testing in an air, all dimples had transformed into parabolic type,

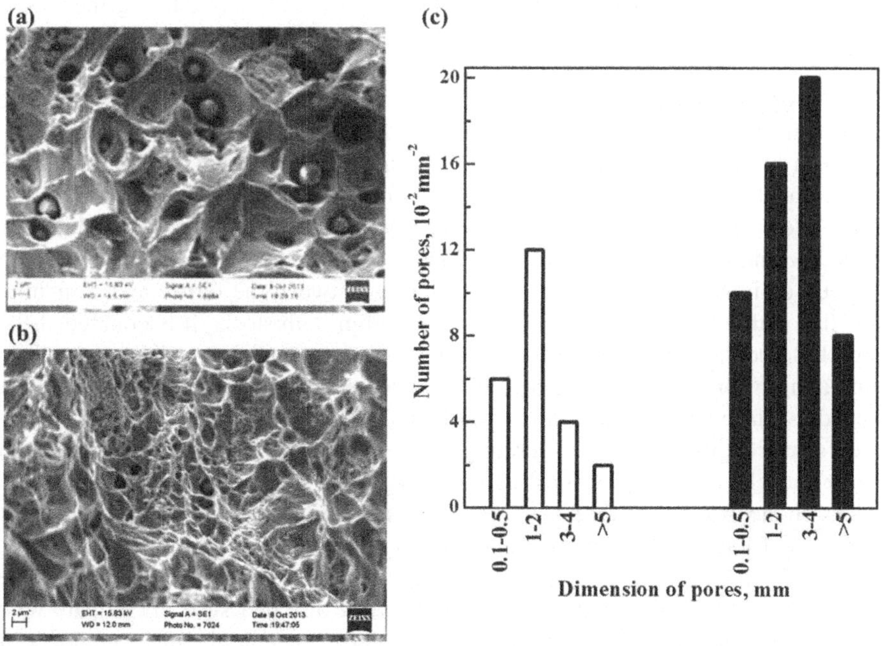

Fig. 4.20 The typical microfractograms obtained in the central parts of the tension specimens **a** from the old and **b** the repair steels tested in air and **c** the corresponding distributions of the pore sizes on the fracture surfaces of the repair (white) and the old (black columns) steels from elements of Shukhov's Tower

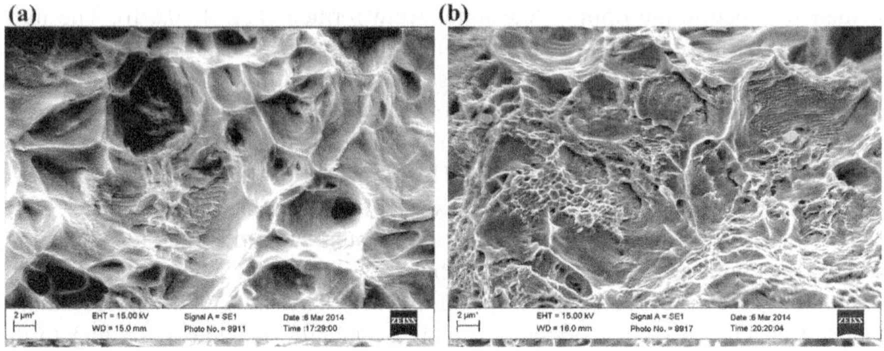

Fig. 4.21 Microfractograms of the **a** old and **b** repair steels from the central parts of the fracture surfaces of the specimens after their tensile tests in air

which are typical for the conical part of the fracture surfaces of smooth specimens and corresponded by the final stage of their destruction due to shear deformation.

During testing of the smooth specimens from old and repair steels in the artifi-cial acid rain environment, the favourable conditions for the cracks initiation were arisen not only in the central parts of their cross sections (as observed during the test in air) but moreover in several points located along their perimeter (Fig. 4.22a, b). As a rule, the destroying of the specimens started from their external surfaces in several places along perimeter in the zones with the predominant location of the perlite grains. In another zones on the fracture surfaces near side surface of specimens (between points of the nucleation of failure under the influence of the corrosion environment) the fractographic signs did not differ from the obtained during testing in air. The propagation of the fracture process was occurred along the boundaries of the ferrite and the perlite grains. The secondary cracks along these boundaries with the well-delimited of the interphase boundaries between the ferrite and cementite lamellas within the perlite grains were observed in both steels. It is concluded the interphase boundaries are most inclined to the influence of the model environment simulating of the effect of precipitation. The appearance of such type damages on the side surface of specimens was contributed to the acceleration of the localizing of deformation processes in the certain cross section along their working length and thereby accelerated their final destruction. It is possible that network of the corrosion cracks on the external surface of the specimens can be responsible for a weak effect of the corrosion environment on the reduction of area of the specimens (Fig. 4.18). The large numbers of the revealed sites of initiation of corrosion cracks from side surface of the specimens is primarily characteristic of repair steel in compare with old one. It is due to higher carbon content and, correspondingly, more numbers of the clusters from the perlite grains in the repair steel. The localization of deformation in one of the cross section of specimen became more possible due to the presence of such cracks on their side surface. It means that deformation degree in this cross section would be less and correspondently the value of the reduction of area of the specimens would be also less than during test in air.

The effect of the corrosion environment was also manifested in the central part of the fracture surfaces of specimens from both steels by presence of fracto-graphic elements, which were considered as the signs of scattered damages inside steels occurred during their exploitation (Fig. 4.22c, d). Almost round holes remaining from the non-metallic inclusions and extending into the deep layers below the fracture surface of the specimens were revealed in both steels. Their number and distribution in the cross section of specimens corresponded to observe during the test in air. But the non-metallic inclusions are no longer visualized in the most of these holes. Moreover, this feature was more pronounced in the old steel and considered as a manifestation of the joint influence of the hydrogenation effect of environment and tensile stresses during specimens testing. The hydrogen absorbed by the metal (both during operation and during the SCC with a SSRT of the specimens) was accumulated in the cavities along the interface boundaries between non-metallic inclusions and ferrite matrix. The hydrogen pressure together with tensile stresses, contributed to the final separation of the non-metallic inclusions with the matrix. As a consequence, the quite large and long cavities were formed in the metal. Thus, the removal of non-metallic inclusions from the fracture surfaces during the final destruction of the specimens was become practically unimpeded. Moreover, the formed network of holes inside the exploited steels was undoubtedly reduced their strength (according to the results of mechanical tests, this decrease can be almost 20%).

However, a more important manifestation of the hydrogenation effect of both exploited steels was the rounded brittle fragments on their fracture surfaces clearly defined in the background of mostly ductile relief covered by the dimples (similar to obtained in air). It is suggested that the damaging along the boundaries between of the ferrite and pearlite grains most likely are initiated at the stage of the long-time operation of the both steels, and hydrogenation during slow loading of specimens in the synthetic acid rain environment facilitated cracking along them. The brittle fragments were revealed in the central normally oriented part of the fracture surfaces of specimens (Fig. 4.22c–e). Their numbers and area on the fracture surface of the repair steel were more than in the old one. In the repair steel, these fragments consisted from the conglomerates of pearlite grains, separated along their interface from the ferrite matrix. These interfaces were considered as the places of the fracture initiations in the inner layers of the specimens. The ferrite bridges between them were more often destroyed by plastic stretching until their rupture (Fig. 4.22c), or sometimes by formation of the typical cleavage facets between neighbouring perlite grains already separated from ferrite grains under hydrogen effect (Fig. 4.22e). In the latter case, the fracture surface of the specimen was more like to the structure of repair steel. In any case, the similar cleavage elements were not observed at the fracture surface of this steel after it testing in air. It enable to associate of their origination with the embrittlement caused by hydro-gen absorbed by the metal during long-term operation whose action is enhanced by the hydrogen absorbed by the metal during the slow deformation of the specimen in the acid rain environment.

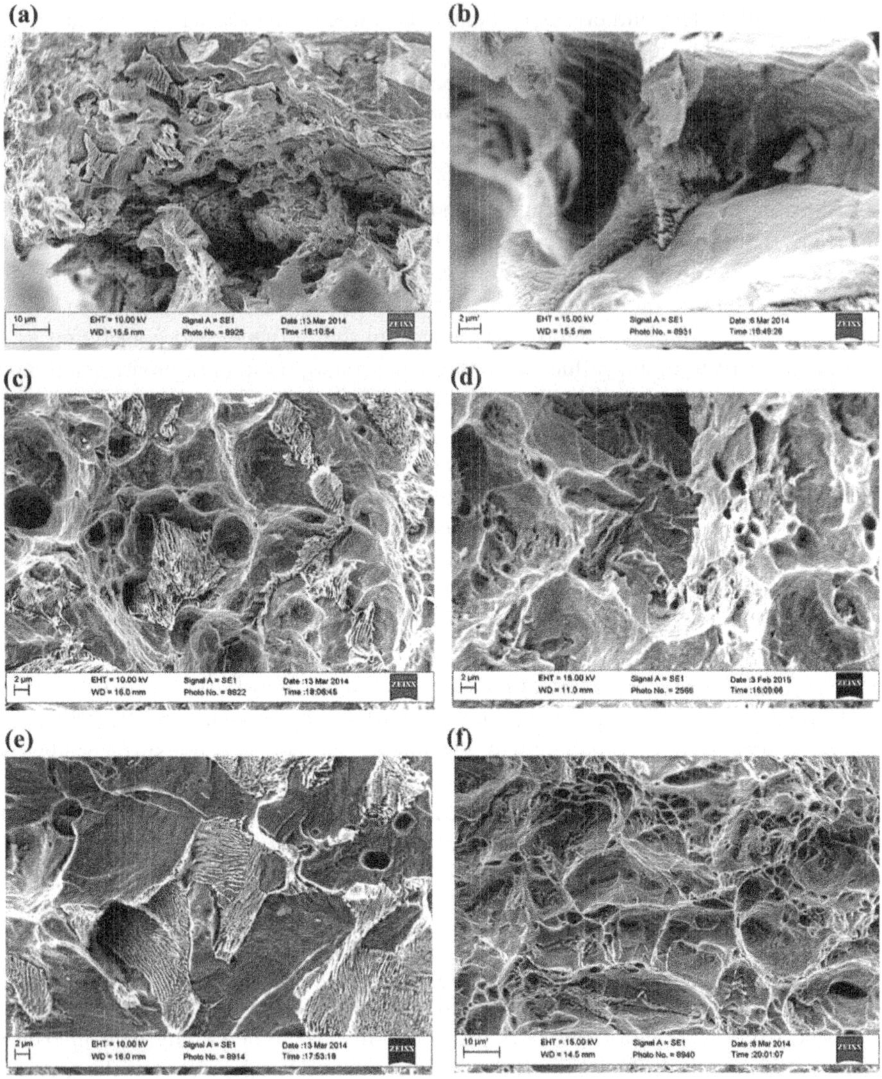

Fig. 4.22 Microfractograms **a**, **c**, **e** of the repair and **b**, **d**, **f** the old steels obtained in the vicinity of the fracture nucleation **a**, **b** from the side surface of specimens and **c**–**f** in the central part of their fracture surfaces after testing on the SCC at SSRT in a synthetic acid rain environment

In the old steel, the embrittlement under hydrogen effect was manifested in the form of small rounded fragments of the brittle fracture with a typical relief of the separation crests in the form of the fan diverged from the points of cleavage initiation with the formation of river patterns on the transgranular facets of the fracture surface. (Fig. 4.22d). The relief around such cleavage fragments remains the ductile (by dimple forming) as in air tests. However, the morphology of the dimples changed under the influence of hydrogen absorbed by the metal. Just about equiaxial and deep dimples formed due to a fairly large stretching of the bridges between them were observed on the fracture surface of specimen tested in the air. However the significant part of the dimples had an elongated shape after the test in the hydrogenation environment (Fig. 4.22f). Moreover, they were more flat (less deep) and on their bottom, as a rule, chains from traces remained by the inclusions were observed. It was assumed that these pores at the bottom of the dimples were served as storage for the hydrogen promoting the growth of such disk-like dimples. The similar mechanism of hydrogen influence on the pore formation inside the metal was firstly suggested by Murakami [45].

Basing on the confirmed assumption about developed damages network in the form of the cleavage fragments (in the old steel) and practically transgranular fracture through the perlite grains and ferrite ones (in repair steel) can explained the higher elongation of specimens of old steel tested in the hydrogenation environment compared with the repair one. All these cleavage elements testify to intensive opening of defects already at the initial stage of the active loading of the specimens at the stage of the uniform plastic deformation along whole their working length. Since such brittle fragments were more frequently observed in the repair steel (because of the higher carbon content in it and, as a consequence, its higher sensibility to hydrogen embrittlement), the elongation of the specimens from this steel was less compared with the old steel.

Since described embrittlement elements were not observed in the central part of the specimens tested in air, their visualization was caused to diffusion of hydrogen from the test environment to the center of the cross section of specimens with following accumulation into already existing defects inside degraded steels. Based on the foregoing, it was believed that hydrogen could facilitate the destruction of metal in the vicinity of local places of the interphase defects concentration formed during long-time operation of steels. And accounting of this aspect of the problem of the steels degradation after long-term exploitation at the ancient objects worthy of the preservation for future generations is important to ensure their integrity.

References

1. Shukhov hyperboloid of rotation. https://sohabr.net/habr/post/391819/. Last accessed 30 Aug 2018
2. Kobe Port Tower. https://en.wikipedia.org/wiki/Kobe_Port_Tower. Last accessed 30 Aug 2018
3. TV towers Guangzhou in China. https://poradu.pp.ua/dim/29940-televezha-guanchzhou-kitay.html. Last accessed 30 Aug 2018
4. The Skyscraper Center. https://www.skyscrapercenter.com/building/aspire-tower/535. Last accessed 30 Aug 2018
5. Graefe R, Suchov VG (1990) 1853–1939 – Die Kunst der sparsamen Konstruktion, DVA, Stuttgart
6. Kutnyi A, Becker K (2011) Raffinierter noch als der Eiffelturm. Deutsch. Ing. Blatt 12:38–39

7. Nykyforchyn HM, Kutnyi AO, Kremin' TD, Tsyrul'nyk OT (2013) Condition diagnostics for the Shukhov hyperbolic gridshell towers in service for over 100 years. Mashynoznavstvo, 1–2, 15–19. (in Ukrainian)
8. Suslova OY, Dushkevych KN, Kovenskyy RC, Pshenizyna AV (2009) The world-famous buildings of V.H. Shukhov in danger. AMIT 4(9), 1–7
9. Adzhigol lighthouse https://vseua.info/adzhigol-s-kij-mayak/. Last accessed 30 Aug 2018
10. Photo Stanislav lighthouse, https://upload.wikimedia.org/wikipedia/commons/4/4b/Stanislav_lighthouse.jpg/. Last accessed 31 Aug 2018
11. Nykyforchyn HM, Tsyrul'nyk OT, Petryna D Yu, Hredil MI (2009) Degradation of steels used in gas main pipelines during their 40-year operation. Strength Mater 41(5):501–505
12. Maruschak P, Danyliuk I, Prentkovskis O, Bishchak R, Pylypenko A, Sorochak A (2014) Degradation of the main gas pipeline material and mechanisms of its fracture. J Civil Eng Manage 20(6):864–872
13. Lesiuk G, Szata M (2014) Fatigue properties and fatigue crack growth in puddled steel with consideration of microstructural degradation processes after 100-years operating time. Proc Eng 74:64–67
14. Lesiuk G, Szata M (2011) Aspects of structural degradation in steels of old bridges by means of fatigue crack propagation. Mater Sci 47:82–88
15. Student OZ, Krechkovs'ka HV (2012) Anisotropy of the mechanical properties of degraded 15Kh1M1F steel after its operation in steam pipelines of thermal power plants. Mater Sci 47(2):590–597
16. Krechkovs'ka GV (2008) Structural changes in exploitation of steam power plant pipeline 15Kh1M1F-type steel concerning with shut downs of power units. Metallofiz Noveishie Tekhnol 30:701–711
17. Student OZ, Krechkovs'ka HV, Palashchuk TE, Hladkyi Ya M (2018) Influence of the long-term operation of 12Kh1MF steel of the bends of main steam pipelines of thermal power plants on its mechanical properties. Mater Sci 53(4):460–467
18. Babii LO, Student OZ, Zagorski A, Markov AD (2007) Creep of degraded 2.25 Cr-Mo steel in hydrogen. Mater Sci 43(5), 701–707
19. Embrittlement of the steel of an oil-trunk pipeline/Tsyrul'nyk OT, Nykyforchyn HM, Zvirko OI, Petryna D Yu (2004) Mater Sci 40 (2), 302–304
20. Nykyforchyn HM, Krechkovs'ka HV, Kutnyi AI, Student OZ (2015) Analysis of long-term in-service degradation of the shukhov tower elements. Strength Mater 47(5):679–688
21. Krechkovs'ka HV, Student OZ, Kutnyi AI, Nykyforchyn HM, Sydor P Ya (2015) Brittle-fracture resistance of the metal of hyperboloid GridshellShukhov tower. Mater Sci 50(4):578–584
22. Student O, Krechkovska H, Lesiuk G, Correia J (2018) Features of the microstructural and mechanical degradation of long term operated mild steel. Int J Struct Integr 9(3):296–306
23. Nykyforchyn HM, Student OZ, Dzioba IR, Stepanyuk SM, Markov AD, Onyshchak YD (2004) Degradation of welded joints of steam pipelines of thermal electric power plants in hydrogenating media. Mater Sci 40(6):836–843
24. Kremen T (2006) The stream of life water. To the 100th anniversary of the founding of the water supply pipeline for Mykolaiv City (1906–2006), Publishing House "Ilion", Mykolaiv. (in Ukrainian)
25. Nykyforchyn HM, Student OZ, Markov AD (2007) Abnormal behavior of high-temperature degradation of the weld metal of low-alloy steel welded joints. Mater Sci 43(1):77–84
26. Kryzhanivs'kyi EI, Nykyforchyn HM (2011) Specific features of hydrogen-induced corrosion degradation of steels of gas and oil pipelines and oil storage reservoirs. Mater Sci 47(2):127–138
27. Nykyforchyn HM, Student OZ, Krechkovs'ka HV, Markov AD (2010) Evaluation of the influence of shutdowns of a technological process on changes in the in-service state of the metal of main steam pipelines of thermal power plants. Mater Sci 46(2):177–189
28. Gabetta G, Nykyforchyn HM, Lunarska E, Vuherer T (2008) In-service degradation of gas trunk pipeline X52 steel. Mater Sci 44(1):104–119

29. Tsyrul'nyk OT, Voloshyn VA, Petryna D Yu, Zvirko OI (2011) Degradation of properties of the metal of welded joints in operating gas mains. Mater Sci 45(5):573–582
30. Starchak VH, Oleksiyenko CO, Ivanenko KM, Tsybulya CD (2008) The danger of the influence of the non-metallic inclusions on the hydrogen of steel degradation Visnyk. YMT 1(1):122–141 (in Ukrainian)
31. Gainer LJ, Wall Work GR (1979) The effect of nonmetallic inclusions on the pitting of mild steel. Corrosion 35(10):435–443
32. Baker MA, Castle JE (1993) The initiation of pitting corrosion at MnS inclusions. Corros Sci 34(4):667–682
33. Ray GP, Jarman RA, Thomas JGN (1985) The influence of non-metallic inclusions on the corrosion fatigue of mild steel. Corros Sci 25(3):171–184
34. Zakharchuk V, Tsyrul'nyk O, Nykyforchyn H (2005) Electrochemical properties of steels in a model hydrogen galvanic couple. Mater Sci 41(2):223–229
35. Nykyforchyn HM, Kurzydlowski K-J, Lunarska E (2008) Hydrogen degradation of steels in long term service conditions. In: Shipilov S (ed) Environment-induced cracking of materials, vol. 2: prediction, industrial developments and evaluations, pp 349–361. Elsevier, Amsterdam.
36. Tsyrulnyk OT, Slobodyan ZV, Zvirko OI, Hredil' MI, Nykyforchyn HM, Gabetta G (2008) Influence of operation of Kh52 steel on corrosion processes in a model solution of gas condensate. Mater Sci 44(5):619–629
37. Voloshyn VA, Zvirko OI, Sydor PY (2015) Influence of the compositions of neutral soil media on the corrosion cracking of pipe steel. Mater Sci 50(5):671–675
38. Sidorenko VM, Fedorov VV, Barabash LV, Pokhmurskii VI (1978) Acceleration of self-diffusion processes in metals under the influence of dissolved hydrogen. Mater Sci 3(6):607–610
39. Archakov YuI, Dobrotvorskii AM, Pokhmurs'kyi VI, Fedorov VV (1995) Mechanism of the influence of hydrogen on the self-diffusion of nickel. Mater Sci 31(4):477–484
40. Goltsov VA (1997) Fundamentals of hydrogen treatment of materials and its classification. Int J Hydrogen Energy 2(3):119–124
41. Krasovskii AY, Lokhman IV, Orynyak IV (2012) Stress-corrosion failures of main pipelines. Strength Mater 44(2):129–143
42. Kryzhanivs'kyi EI, Hrabovs'kyi RS, Mandryk OM (2014) Estimation of the serviceability of oil and gas pipelines after long-term operation according to the parameters of their defectiveness. Mater Sci 49(1):117–123
43. Nykyforchyn H (2013) Environmentally assisted degradation of the physical and mechanical properties of long-term exploited structural steels. Newsletter 52(2):20–24
44. Tsyrul'nyk OT (2014) Application of the electrochemical methods in the diagnostics of the engineering state of structural materials. Mater Sci 49(4):449–460
45. Murakami Y (2008) Effect of hydrogen on fatigue crack growns of metals. In: 17th European conference on fracture, , pp 25–41. 2–5 September, 2008, Brno

Chapter 5
Fatigue and Fracture Behaviour of Long Term Operated Bridge Materials and Components

Abstract This book chapter intends to provide a fatigue database for some representative materials of old Portuguese metallic bridges (Eiffel, Luiz I, Fão, Pinhão and Trezói). Puddle irons and structural steels are covered, depending of construction year of the bridge. Besides the base materials, also S–N fatigue data of riveted joints made of original rivets or new rivets are presented, but in any case using original plates. As regards the materials characterization, S–N fatigue data obtained using smooth specimens is presented; also, fatigue crack propagation data obtained using CT specimens is presented. The monotonic tensile properties are also characterized as well as the microstructures. S–N fatigue data for riveted joints was also presented and compared with existing literature data, including existing design curves. This comparison shows some data falling below the design S–N curves that corroborates the presence of cracks originated by the prior operation of the bridge or material degradation. Regarding the strain-life behaviour of the materials investigated, the number of transition reversals decreases with increasing age of the materials. Only the material from the Trezói bridge shows a transition life typical of low carbon steels. The fatigue crack propagation tests showed that Paris law gives a good description of the fatigue crack growth data with an exponent always greater than 3. The *C* coefficient was in same range of that recommended in literature for modern construction steels.

5.1 Introduction

The maintenance and safety of existing bridges is a major concern of governmental agencies. In particular, the safety of old metallic riveted road and railway bridges fabricated and put into service at the end of the 19th century and beginning of 20th century deserve particular attention, since they were designed taking into account traffic conditions, both in terms of vehicle gross weight and frequency, completely different from those arising nowadays. Also, the current design procedures were not yet fully developed or even did not exist in the 19th century and design engineers were not aware of some important phenomena, such as fatigue. Nevertheless, fatigue was only intensively studied in the 20th century. Fatigue failures are a concern for steel

G. Lesiuk et al., *Degradation Theory of Long Term Operated Materials and Structures*, Structural Integrity 15, https://doi.org/10.1007/978-3-030-43710-7_5

bridges due to the likelihood of the steel to deteriorate under variable stresses [1–3], being recognized as the major cause of failure in metallic bridges [4]. In order to assure high safety levels in old riveted metallic bridges, road and railway authorities have to invest heavily in their maintenance and retrofitting.

Residual life calculations of existing bridges in operation should take into account fatigue as a progressive damaging mechanism. A consistent residual life prediction should be based on actual fatigue data from bridge members being assessed.

The S–N approach is widely used to assess the fatigue damage of riveted connections, which is included in design codes of practice (ex: EC3-1-9 [5], AASHTO [6]). This approach is based on detail category S–N curves and relates the total number of cycles to failure to the applied stress range. Alternatively, Fracture Mechanics has been applied to assess the residual fatigue life of damaged riveted connections [7]. This approach requires the knowledge of the initial defect, which may be assessed by inspection. In order to turn the Fracture Mechanics in a more versatile design alternative than the S–N approach, it should be complemented by another approach to assess the crack initiation [8, 9]. Local approaches to fatigue, based on local or notch stresses or strains, are frequently used to assess the fatigue crack initiation [10]. The application of the local approaches to fatigue as well as the Fracture Mechanics require the knowledge of experimental fatigue data of the plain material, such as strain-life fatigue data as well as fatigue crack propagation data. This data is generally available in the literature for modern construction steels. However, there is a lack of experimental fatigue data for materials used in old metallic riveted bridges. Many of these materials are puddle irons, precursors of modern construction steels, which are not sufficiently studied.

This book chapter aims at the characterization of the fatigue behaviour of different materials from a representative group of Portuguese old metallic riveted bridges, namely the Eiffel, Luiz I, Fão, Pinhão and Trezói bridges. Besides the strain-life and the fatigue crack propagation behaviour the monotonic and cyclic elastoplastic behaviours, the microstructures and the chemical composition of the materials are also characterized [11–18]. In addition to the material characterization, S–N fatigue data from riveted joints is also presented in this chapter. A review of data derived by the authors and other data dispersed in the literature is gathered and compiled in this chapter.

A brief description of the riveted bridges from which materials have been characterized, is presented in the next paragraphs. The oldest bridge is the Eiffel bridge that was designed by Gustave Eiffel and inaugurated on 30th of June 1878 (see Fig. 5.1). This 573 m long and 6 m wide bridge, made of a continuous deck composed by nine spans, crosses the Lima river, between Darque and Viana do Castelo, and serves both road and railway traffic. The material used in the experimental program was extracted from a beam of the railway Darque viaduct, which was removed during the last rehabilitation, occurred between 1st February 2006 and 30st October 2007 [11, 18].

The second oldest bridge is the Luiz I bridge (see Fig. 5.2) that was also designed by Gustave Eiffel and commissioned in 31st October 1886. This bridge crosses the Douro river and links the Porto and Gaia cities. The main features of the Luiz I bridge

Fig. 5.1 Riveted metallic Eiffel bridge in Viana do Castelo

Fig. 5.2 Riveted metallic Luiz I bridge in Porto

are: double deck supported by an arch; span of 172 m; arch radius of 45 m; length of upper deck is 391.25 m; length of lower deck is 174 m and width is 8 m (original design was 6 m). A diagonal member 1600 mm in length was removed from the Luiz I bridge, that was also rehabilitated some years ago [12, 18].

The third oldest bridge included in this study is the Fão road bridge. It was designed by Abel Maria Mota, under the supervision of Reynau, at the end of 19th century and was inaugurated on 7th of August 1892. This bridge, made of a continuous deck with eight spans of 33.5 m each, supported on masonry piers, crosses the Cávado river at Esposende, as illustrated in Fig. 5.3, it was rehabilitated about 11 years ago and 7 diagonals were replaced. The removed diagonals were used in the fatigue experimental program [13, 14, 18].

The fourth oldest bridge considered in this research, illustrated in Fig. 5.4, is the Pinhão road bridge, also designed by Gustave Eiffel at the end of 19th century and built between 1903 and 1906. This bridge crosses the Douro river at Pinhão. Consisting in three spans of 68.8 m each and one span of 10 m; there is only one deck 6 m wide, divided into one traffic lane 4.60 m wide and two sidewalks 0.675 m

Fig. 5.3 Riveted metallic Fão bridge in Esposende

Fig. 5.4 Riveted metallic Pinhão bridge

wide each. A diagonal 1500 mm in length and a bracing 1400 mm in length were removed and replaced by new material [15, 18]. The bridge was also rehabilitated some years ago.

Finally, the youngest bridge included in this study is the Trezói railway bridge, illustrated in Fig. 5.5. This bridge makes part of the Beira Alta railway line and was inaugurated on 20th of August 1956. The deck is composed of three continuous spans of 39, 48 and 39 m, totalizing a bridge length of 126 m. The deck width is 4.40 m throughout the bridge length. The Trezói bridge has about half age of above four referred bridges, which in turn are more than 100 years old. A 3000 mm long bracing was removed from the Trezói bridge only for research purposes, since this bridge was not recently subject to any rehabilitation work [16, 18]. While the material used in the Trezói bridge is a construction steel, the materials used in the older bridges are very likely puddle irons. At the end of the XIX century and beginning of the XX century, the qualities of the puddle irons changed appreciably.

Fig. 5.5 Riveted metallic Trezói bridge

5.2 Monotonic Tensile Strength Properties Characterization

All materials from the referred structures were characterized using monotonic tensile tests. According to the Portuguese NP 10002-1 standard [19], round specimens machined from original members removed from the five bridges under investigation, exhibiting the dimensions listed in the Table 5.1, were subjected to monotonic increasing loading. Distinct diameters were selected since the materials samples extracted from the bridges showed distinct sizes. Besides the strength properties, the elastic properties of the materials were estimated, in particular the Young modulus and Poisson ratio.

Eight specimens were extracted from a south viaduct girder of the Eiffel bridge, in particular from the angle connecting the web and the flange (TC specimens) and from the web (TA specimens). Five specimens were extracted from a diagonal of the Luiz I bridge; twenty-two specimens were extracted from side diagonals of the Fão bridge;

Table 5.1 Cross-sections of the specimens used in the monotonic tensile tests of the materials from the different bridges

Bridge material		Diameter mm	Cross-section mm^2
Eiffel	Viaduct	4	12.57
	Viaduct	5	19.63
	Viaduct	6	28.27
Luiz I	Diagonal	6	28.27
	Diagonal	8	50.27
Fão	Diagonal	6	28.27
Pinhão	Diagonal	5	19.63
	Bracing	8	50.27
Trezói	Bracing	8	50.27

fourteen specimens were machined from a diagonal (CT specimens) and a bracing (BT specimens) from the Pinhão bridge; finally, three specimens of the material from Trezói bridge were prepared from the removed upper bracing. Figure 5.6 presents a photo of a typical tensile specimen.

The results from the monotonic tensile tests, for the materials under consideration of the Eiffel, Luiz I, Fão, Pinhão and Trezói bridges are presented in Tables 5.2, 5.3, 5.4, 5.5 and 5.6, respectively, where f_u is the ultimate tensile strength, f_y is the higher yield stress, A is the elongation at fracture and Z is the reduction in cross-section at failure. Tables 5.7, 5.8, 5.9, 5.10 and 5.11 show the mean values, standard deviation and coefficient of variation (COV) of the various properties evaluated with the monotonic tensile tests for the materials of the Eiffel, Luiz I, Fão and Trezói bridges, respectively.

Table 5.7 shows the mean values, standard deviation and COV of the properties for the materials extracted from the angle and web of the south viaduct girder of the Eiffel bridge, respectively the TA and TC specimens. The table also presents global mean, standard deviation and COV of the properties for all specimens taken from the web and the upper angle of the south viaduct girder of the Eiffel bridge.

The mean tensile, standard deviation and COV of the properties of the material extracted from the diagonal and bracing of the Pinhão bridge, referred as CT and BT specimens, respectively, are shown in Table 5.10. Global mean, standard deviation and COV of the properties for all specimens, analysed together, taken from the diagonal and bracing of the Pinhão bridge, are also included in Table 5.10.

Fig. 5.6 Typical specimen used in monotonic tensile tests [16]

Table 5.2 Summary of results from the monotonic tensile tests for the material from Eiffel bridge [11]

Specimen	f_u MPa	f_y MPa	A %	Z %
TA1	387.0	345.0	6.80	4.70
TA2	376.0	309.0	13.20	14.70
TA3	285.0	282.0	2.00	0.80
TA4	276.0	275.0	0.80	1.60
TC1	424.0	297.0	14.30	35.30
TC2	428.0	295.0	22.00	31.70
TC3	286.0	269.0	3.20	2.00
TC4	272.0	267.0	2.80	2.00

Table 5.3 Summary of results from the monotonic tensile tests for the material from Luiz I bridge [12]

Specimen	f_uMPa	f_yMPa	A %	Z %
S1T1	435.0	336.0	14.00	11.00
S2T1	374.0	275.0	13.60	9.70
S3T1	395.0	302.0	29.30	39.20
S3T2	383.0	287.0	25.30	37.70
S3T3	396.0	313.0	23.80	38.30

Table 5.4 Summary of results from the monotonic tensile tests for the material from Fão bridge [13]

Specimen	f_uMPa	f_yMPa	A %	Z %
1	345.5	209.4	23.43	8.31
2	337.2	198.5	–	12.06
3				
4	423.9	288.8	21.60	13.07
5	317.8	208.0	12.77	7.64
6	425.0	276.8	25.87	13.87
7	371.3	249.0	–	12.81
8	–	–	–	–
9	333.8	225.6	–	8.79
10	355.5	216.1	27.00	14.86
11	342.7	195.2	21.47	13.26
12	348.6	223.8	12.97	7.78
13	354.8	210.7	19.03	14.52
14	357.8	239.3	24.07	12.66
15	378.3	237.3	21.67	20.77
16	337.3	200.7	–	15.05
17	396.7	245.6	22.23	17.91
18	352.7	210.2	23.27	16.07
19	369.0	226.4	24.77	18.07
20	343.0	196.2	21.43	9.81
21	361.9	232.7	–	14.23
22	333.8	107.7	45.33	9.58

Clearly the material from Trezói bridge exhibits the highest strength properties associated with high ductility. The material from the Eiffel bridge—the oldest one—shows the lowest ductility properties. The material from the Pinhão bridge shows the highest ductility properties. When compared with the materials from the other centenary bridges, the material from the Pinhão bridge shows a very good strength/ductility

Table 5.5 Summary of results from the monotonic tensile tests for the material from Pinhão bridge [15]

Specimen	f_uMPa	f_yMPa	A %	Z %
CT1	372.0	297.0	31.30	69.40
CT2	373.0	274.0	–	69.40
CT3	365.0	289.0	–	72.00
CT4	365.0	275.0	36.00	71.40
CT5	367.3	278.8	34.00	69.80
CT6	365.8	302.1	34.80	69.80
CT7	364.2	274.1	29.00	68.40
BT1	377.0	377.0	–	77.00
BT2	361.0	361.0	–	71.00
BT3	348.0	344.0	–	73.00
BT4	354.0	352.0	–	71.00
BT5	345.0	263.7	33.20	75.00
BT6	348.7	300.7	33.60	68.00
BT7	348.9	294.1	33.60	68.40

Table 5.6 Summary of results from the monotonic tensile tests for the material from Trezói bridge [16, 17]

Specimen	f_uMPa	f_yMPa	A %	Z %
T1	455.0	392.0	22.00	66.00
T2	501.0	401.0	23.00	71.00
T3	464.0	402.0	24.00	62.00

Table 5.7 Mean, standard deviation and COV of the monotonic tensile properties for the material from Eiffel bridge

Specimen	Statistical data	f_uMPa	f_yMPa	A %	Z %
TA	Mean	331.0	302.8	5.70	5.45
	Standard deviation	58.6	31.8	5.63	6.39
	COV [%]	17.7	10.5	98.81	117.28
TC	Mean	352.5	282.0	10.58	17.75
	Standard deviation	85.1	16.2	9.30	18.25
	COV [%]	24.1	5.7	87.91	102.79
TA/TC	Mean	341.8	292.4	8.14	11.60
	Standard deviation	68.6	25.8	7.58	14.26
	COV [%]	20.1	8.8	93.12	122.95

Table 5.8 Mean, standard deviation and COV of the monotonic tensile properties for the material from Luiz I bridge

Statistical data	f_uMPa	f_yMPa	A %	Z %
Mean	396.6	302.6	21.20	27.18
Standard deviation	23.3	23.6	7.05	15.38
COV [%]	5.9	7.8	33.25	56.58

Table 5.9 Mean, standard deviation and COV of the monotonic tensile properties for the material from Fão bridge [13]

Statistical data	f_uMPa	f_yMPa	A %	Z %
Mean	359.3	219.9	23.13	13.06
Standard deviation	28.5	36.6	7.35	3.63
COV [%]	7.9	16.6	31.80	27.81

Table 5.10 Mean, standard deviation and COV of the monotonic tensile properties for the material from Pinhão bridge

Specimen	Statistical data	f_uMPa	f_yMPa	A %	Z %
CT	Mean	367.5	284.3	33.02	70.03
	Standard deviation	3.6	11.7	2.83	1.25
	COV [%]	1.0	4.1	8.58	1.78
BT	Mean	354.7	327.5	33.47	71.91
	Standard deviation	11.2	41.5	0.23	3.31
	COV [%]	3.1	12.7	0.69	4.61
CT/BT	Mean	361.1	305.9	33.19	70.97
	Standard deviation	10.4	36.9	2.16	2.60
	COV [%]	2.9	12.1	6.50	3.66

Table 5.11 Mean, standard deviation and COV of the monotonic tensile properties for the material from Trezói bridge

Statistical data	f_uMPa	f_yMPa	A %	Z %
Mean	473.3	398.3	23.00	66.33
Standard deviation	24.4	5.5	1.00	4.51
COV [%]	5.2	1.4	4.35	6.80

ratio [17]. It is worthwhile mentioning that the Pinhão bridge is the youngest of the centenary bridges and at that time the steel production processes evolved quickly in quality. The material from the Pinhão bridge shows characteristics very similar to the modern construction steels.

Table 5.12 Strain hardening coefficients and exponents for the materials from the bridges

Bridge materials	*K* MPa	*n*
Eiffel	411.07	0.0585
Luiz I	506.20	0.0902
Fão	477.53	0.1300
Pinhão	444.29	0.1166
Trezói	586.86	0.0957

The monotonic stress-strain curves were further correlated each other using the Ramberg-Osgood relation:

$$\varepsilon = \frac{\sigma}{E} + \left(\frac{\sigma}{K}\right)^{1/n} \tag{5.1}$$

where ε and σ are, respectively, the strain and stress, E is the Young modulus, K and n are the strain hardening coefficient and exponent. The resulting monotonic properties (K and n) are summarized in Table 5.12 for all materials under consideration.

The elastic properties, represented by the Young modulus, E, and Poisson ratio, ν, of the materials from the Luiz I and Fão bridges were computed directly from strain gauge measurements (see Fig. 5.7). The elastic properties of the materials from the Eiffel and Trezói bridges were computed indirectly from cyclic elastoplastic analysis. For the material of the Trezói bridge, the elastic properties were computed indirectly from analysis of the monotonic tensile tests. Table 5.13 summarizes the estimates of the elastic properties for all materials.

a) b) c)

Fig. 5.7 Specimens used in the assessment of the elasticity modulus and Poisson ratio for the material from the Fão bridge: **a** specimens series; **b** and **c** longitudinal and transverse strain gauge [13]

Table 5.13 Elastic properties of the materials from the bridges

Bridge material	*E* GPa	ν
Eiffel[b]	193.10	0.300
Luiz I[a]	192.70	0.255
Fão[a]	198.70	0.260
Pinhão[c]	210.68	–
Trezói[b]	198.49	0.320

[a]Computed directly from strain gauge measurements
[b]Computed indirectly from cyclic elastoplastic analysis
[c]Computed indirectly from monotonic tensile data

5.3 Metallographic Analysis of the Materials

5.3.1 Optical Microscope Observations

Figure 5.8 illustrates typical microstructures of the materials under investigation. In general, the microstructures strengthen the conclusion that materials are mainly composed of a ferrite matrix. For some cases, namely the material from Pinhão and Trezói bridges, perlite is observed. These latter materials show a more homogeneous microstructure of regular grains than the other materials. All materials exhibit lined-up inclusions induced by deformation, the density of them being however is higher in the centenary materials, thus limiting the material ductility and increasing the scatter of mechanical properties [11–13, 15–18].

5.3.2 Chemical Composition

Table 5.14 summarizes the chemical composition of some material samples from the bridges under investigation [18]. The chemical compositions were assessed using spark emission spectrometry technique. The material from the Trezói bridge exhibits the lowest sulphur and phosphorus contents, which is consistent with the age of the material. The Trezói material is a ferrite structural steel, since it has a very small amount of carbon. The materials from the centenary bridges show significant variable chemical compositions, due to the typical heterogeneous microstructures of these materials. Due to the low carbon content, they should exhibit a ferrite matrix. The samples from Luiz I bridge show higher carbon, silicon and manganese contents than the other materials. This material revealed a significant heterogeneous chemical composition, which is consistent with the heterogeneous microstructure of the puddle irons. The higher carbon content measured is due to local segregations. The higher silicon content may be probably due to the use of Si in the metal desoxidation/desulphurization process.

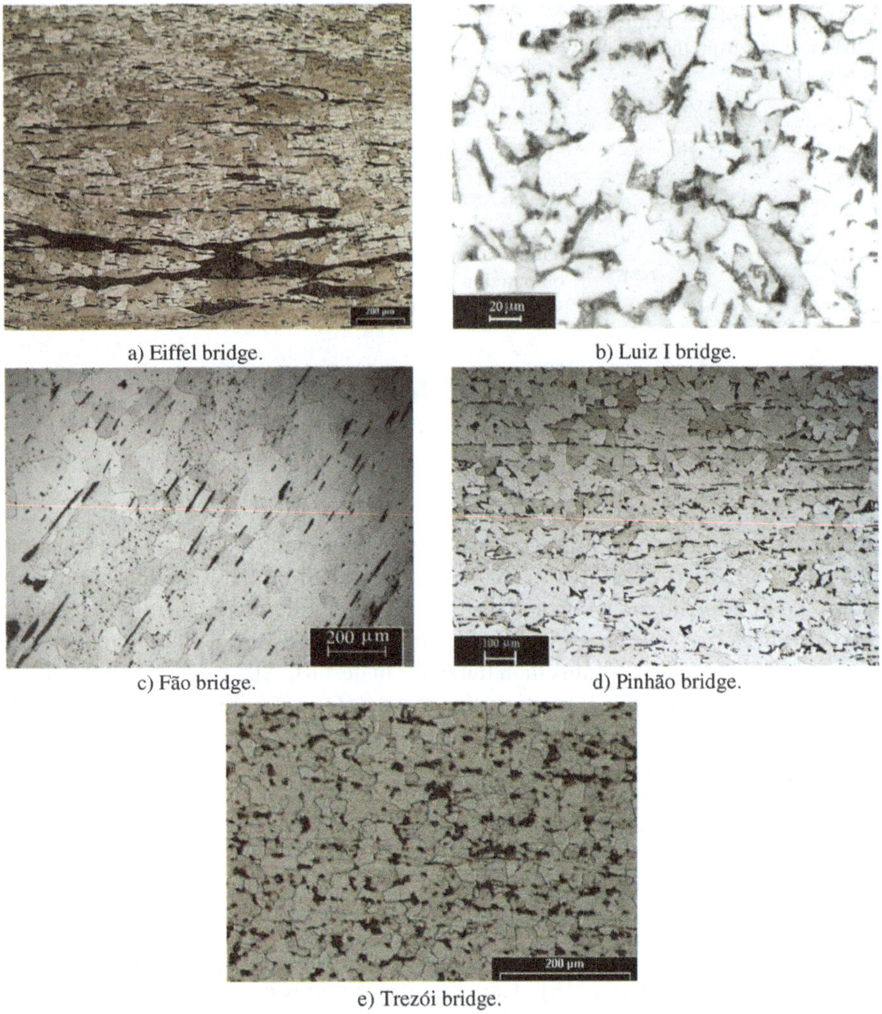

Fig. 5.8 Microstructures of the materials from the bridges

5.4 Fatigue Behaviour Characterization of the Materials

This section presents the strain-life fatigue data obtained for the materials from the bridge under consideration. Besides the fatigue behaviour characterization, the cyclic elastoplastic behaviour of the materials are also presented. Strain-life fatigue data from tests performed on smooth specimens may be used to assess the crack initiation, in the framework of local approaches to fatigue [20].

Smooth specimens with rectangular cross section were machined from material samples from the Eiffel, Luiz I, Fão and Trezói bridges (see Fig. 5.9). These

Table 5.14 Chemical composition for each material of the bridges (wt%) [18]

Brigde	Material	%C	%Si	%Mn	%P	%S
Eiffel	Darque Viaduct bridge*	0.23	0.39	1.78	>0.15	>0.15
		0.81	0.24	2.71	>0.15	>0.15
Luiz I	Diagonal	0.72	0.34	2.09	>0.15	>0.15
Fão	Diagonal	0.09	0.06	0.13	0.14	0.007
Pinhão	Diagonal	0.06	<0.01	0.04	0.04	0.03
	Bracing	0.05	<0.01	0.34	0.04	0.04
Trezói	Bracing	0.06	0.03	0.34	0.02	0.02

*Determined with a portable emission spectrometry

specimens were used to carry out fatigue tests under strain-controlled conditions, according to the ASTM E606 Standard [21]. The specimens were machined according to the rolling (longitudinal) direction of the members removed from the bridges. The dimensions of the specimens were not the same for each material, since the available material thicknesses were not constant. The faces of the specimens were grinded and additionally polished in the central gauge zone.

The tests were performed in an Instron close-loop servo hydraulic machine, model 8801, rated to 100 KN. The deformation was measured using an Instron clip gauge, model 2620-602 with displacement range of ±2.5 mm. A base gauge length of 12.5 mm was used with specimens from the Luiz I and Eiffel bridges; for materials from Eiffel, Fão and Trezói bridges, a base gauge length of 25 mm was chosen. A sinusoidal waveform was taken as command signal. The fatigue tests, in all series, were conducted at room-temperature in air under constant strain amplitudes at a frequency adjusted to result an average strain rate of 0.008/s. However, that average strain rate was not followed in the tests that resulting in a very high number of cycles (high cycle fatigue). In these cases, a higher frequency was selected in order to reduce the testing times.

The choice of specimens with rectangular cross-section instead of a circular one was justified by the fact that the structural elements removed from the bridges have reduced thickness to prepare round specimens. Table 5.15 summarizes the

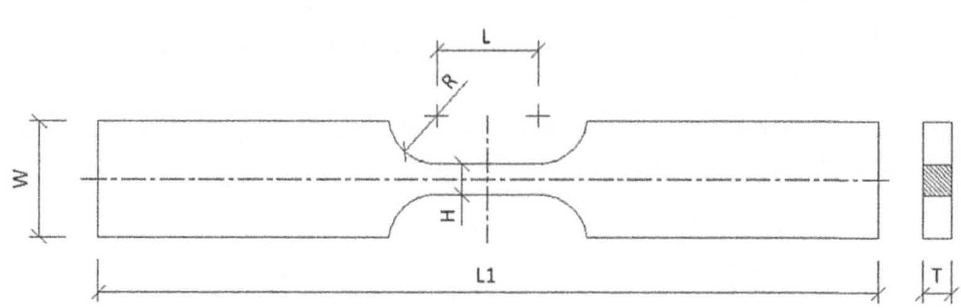

Fig. 5.9 Geometry of the specimens used in fatigue tests under strain-control conditions

Table 5.15 Nominal dimensions of the smooth dog-bone specimens used in fatigue tests

Bridge material	No. of specimens	W mm	T mm	L mm	L1 mm	R mm	H mm
Eiffel	12	22.0	6.0	20.0	150.0	15.0	7.0
	15	20.0	5.0	15.0	150.0	10.0	6.0
Luiz I	15	20.0	5.0	15.0	150.0	10.0	6.0
Fão	35	30.0	7.5	26.0	200.0	12.5	8.0
Trezói	10	30.0	7.5	26.0	200.0	12.5	8.0

nominal dimensions of the smooth dog-bone specimens used in the fatigue tests of the materials from the bridges [13, 17, 18].

The average gauge cross sections were $6.4 \times 5.2\ \text{mm}^2$, $4.9 \times 6.8\ \text{mm}^2$ and $7.5 \times 8.1\ \text{mm}^2$, respectively for the Eiffel (27 specimens), Luiz I (15 specimens) and Trezói (10 specimens) bridges. Regarding the Fão bridge, two test series were prepared with average cross sections of $5.4 \times 7.3\ \text{mm}^2$ and $5.3 \times 7.5\ \text{mm}^2$, which were tested under strain ratios equal to -1 (14 specimens) and 0 (18 specimens), respectively. For the other bridge materials, only one test series was tested, namely under strain ratio equal to -1. Tables 5.10, 5.11 and 5.12 illustrate some specimens tested respectively for the materials from Eiffel, Fão and Trezói bridges.

Tables 5.16, 5.17, 5.18 and 5.19 summarize the program of strain-controlled fatigue tests performed on smooth specimens of materials from Eiffel, Luiz I, Trezói and Fão bridges, respectively. In the referred tables, L_1 and L_2 represent the length of the central section of the smooth specimens, S_t is the cross-section, R_ε the strain ratio, $\Delta\varepsilon$ the strain range, $d\varepsilon/dt$ the average strain rate and f the frequency used for each test.

Tables 5.20, 5.21, 5.22 and 5.23 summarize the results of the test program of smooth specimens for the materials from the Eiffel, Trezói, Luiz I and Fão bridges, respectively. The tables include the total strain range applied to the specimens, $\Delta\varepsilon$, the resulting elastic strain ranges, $\Delta\varepsilon_E$, the plastic strain ranges, $\Delta\varepsilon_P$, the stress range, $\Delta\sigma$, the number of cycles to failure, N_f, and the number of reversals to failure, $2N_f$. The plastic strain ranges and the stress ranges were determined based on the stabilized hysteresis loops. The range of the fully-reversed strains ($R_\varepsilon = -1$) varied, respectively, from 0.166 to 1.00%, 0.20 to 1.00% and 0.20 to 2.50%, for the materials from Eiffel, Luiz I and Trezói bridges. Regarding the material from Fão bridge, the range of applied strains varied, respectively, from 0.20 to 0.80% and 0.15 to 1.50%, for $R_\varepsilon = -1$ and $R_\varepsilon = 0$.

The presented results will be used in the definition of strain-life relations characterizing the fatigue resistance of the materials of the bridges (Figs. 5.10, 5.11 and 5.12).

Table 5.16 Plan of fatigue tests with smooth specimens of material from the Eiffel bridge

Specimens	L_1 mm	L_2 mm	S_t mm^2	R_ε	$\Delta\varepsilon$ %	$d\varepsilon/dt$ %/s	f Hz
LCF-viana-050-01	6.77	5.59	37.84	−1	0.500	0.80	0.800
LCF-viana-050-02	6.79	5.59	37.96		0.500	0.80	0.800
LCF-viana-050-03	6.83	5.64	38.52		0.500	0.80	0.800
LCF-viana-040-01	6.80	5.60	38.08		0.400	0.80	1.000
LCF-viana-040-02	6.81	5.63	38.34		0.400	0.80	1.000
LCF-viana-040-03	6.87	5.63	38.68		0.400	0.80	1.000
LCF-viana-030-01	6.82	5.60	38.19		0.300	0.80	1.333
LCF-viana-030-02	6.76	5.60	37.86		0.300	0.80	1.333
LCF-viana-020-01	6.76	5.61	37.92		0.200	0.80	2.000
LCF-viana-020-02	6.78	5.61	38.04		0.200	0.80	2.000
LCF-viana-075-01	6.82	5.63	38.40		0.750	0.80	0.533
LCF-viana-075-02	6.86	5.61	38.48		0.750	0.80	0.533
LCF_050_01	6.16	4.88	30.0608		0.500	0.80	0.800
LCF_040_01	6.18	5.02	31.0236		0.400	0.80	1.000
LCF_030_01	6.11	4.86	29.6946		0.300	0.80	1.333
LCF_030_02	6.23	4.91	30.5893		0.300	0.80	1.333
LCF_025_01	6.18	4.89	30.2202		0.250	0.80	1.600
LCF_007_01	6.12	4.9	29.988		0.256	–	15.000
LCF_006_01	6.21	5.01	31.1121		0.195	–	15.000
LCF_005_01	6.29	4.93	31.0097		0.166	–	15.000
LCF_0055_01	6.06	4.93	29.8758		0.188	–	15.000
LCF_006_02	5.87	4.9	28.763		0.203	–	15.000
LCF_0055_02	6.05	4.92	29.766		0.188	–	15.000
LCF_060_01	5.95	4.97	29.5715		0.600	0.80	0.667
LCF_070_01	6.09	4.9	29.841		0.700	0.80	0.571
LCF_080_01	6.12	4.98	30.4776		0.800	0.80	0.500
LCF_100_01	5.85	4.92	28.782		1.000	0.80	0.400

5.4.1 Cyclic Elastoplastic Behaviour

Metals when subjected to reversible plastic strains, exhibit a cyclic elastoplastic behaviour which is distinct from its monotonic plastic behaviour. The cyclic elastoplastic behaviour may be characterized by several phenomena, namely: (i) cyclic hardening or cyclic softening; (ii) cyclic mean stress relaxation; (iii) cyclic creep or progressive plastic deformation. In addition to these cyclic behaviours, memory effect and dependency on the strain rate may occur for some materials.

The cyclic curve of the material can be determined connecting the tips of several stabilized hysteresis loops, obtained for different strain amplitudes. The technique

Table 5.17 Plan of fatigue tests with smooth specimens of material from the Luiz I bridge

Specimens	L_l mm	L_2 mm	S_t mm^2	R_ε	$\Delta\varepsilon$ %	$d\varepsilon/dt$ %/s	f Hz
LCF-Luiz1A-050-01	4.95	6.75	33.41	−1	1.000	0.80	0.400
LCF-Luiz1B-050-02	4.95	6.77	33.51		1.000	0.80	0.400
LCF-Luiz1C-0375-01	4.94	6.76	33.39		0.750	0.80	0.533
LCF-Luiz1D-0375-02	4.92	6.80	33.46		0.750	0.80	0.533
LCF-Luiz1E-025-01	4.92	6.76	33.26		0.500	0.80	0.800
LCF-Luiz1F-025-02	4.92	6.79	33.41		0.500	0.80	0.800
LCF-Luiz1G-020-01	4.92	6.79	33.41		0.400	0.80	1.000
LCF-Luiz1H-020-02	4.92	6.77	33.31		0.400	0.80	1.000
LCF-Luiz1I-015-01	4.95	6.79	33.61		0.300	0.80	1.333
LCF-Luiz1J-015-02	4.93	6.74	33.23		0.300	0.80	1.333
LCF-Luiz1K-015-03	4.93	6.75	33.28		0.300	0.80	1.333
LCF-Luiz1L-015-04	4.92	6.75	33.21		0.300	0.80	1.333
LCF-Luiz1M-010-01	4.94	6.78	33.49		0.200	0.80	5.000
LCF-Luiz1N-010-02	4.93	6.81	33.57		0.200	0.80	5.000
LCF-Luiz1O-010-03	4.93	6.77	33.38		0.200	0.80	5.000

Table 5.18 Plan of fatigue tests with smooth specimens of material from the Trezói bridge [17]

Specimens	L_l mm	L_2 mm	S_t mm^2	R_ε	$\Delta\varepsilon$ %	$d\varepsilon/dt$ %/s	f Hz
LCF1	8.00	7.38	59.04	−1	1.000	0.80	0.400
LCF2	7.95	7.48	59.47		0.750	0.80	0.533
LCF3	7.96	7.50	59.70		0.500	0.80	0.800
LCF4	8.08	7.49	60.52		0.300	0.80	1.333
LCF5	8.11	7.53	61.07		0.200	0.80	2.000
LCF6	8.08	7.48	60.44		0.300	0.80	1.333
LCF7	8.04	7.40	59.50		0.500	0.80	0.800
LCF8	8.17	7.44	60.78		1.500	0.80	0.267
LCF9	8.13	7.42	60.32		2.000	0.80	0.200
LCF10	8.29	7.48	62.01		2.500	0.80	0.160

for the evaluation of the cyclic curve may be based on a single specimen test for each point of the cyclic curve, which was the case followed to analyse the data available for the bridge materials. The cyclic curve can be described by the relation proposed by Morrow [22]:

$$\frac{\Delta\sigma}{2} = K'\left(\frac{\Delta\varepsilon^P}{2}\right)^{n'} \tag{5.2}$$

Table 5.19 Plan of fatigue tests of smooth specimens of material from the Fão bridge [13]

Specimens	L_1 mm	L_2 mm	S_t mm^2	R_ε	$\Delta\varepsilon$ %	$d\varepsilon/dt$ %/s	fHz
LCF-e020-01	5.88	7.46	43.86	−1	0.400	0.80	1.000
LCF-e020-02	6.01	7.42	44.59		0.400	0.80	1.000
LCF-e020-03	4.72	7.41	34.98		0.400	0.80	1.000
LCF-e015-01	5.09	7.33	37.31		0.300	0.80	1.333
LCF-e015-02	5.31	7.27	38.60		0.300	0.80	1.333
LCF-e015-03	5.70	7.32	41.72		0.300	0.80	1.333
LCF-e040-01	4.72	7.37	34.79		0.800	0.80	0.500
LCF-e040-02	5.12	7.41	37.94		0.800	0.80	0.500
LCF_e025-01	5.87	7.39	43.38		0.500	0.80	0.800
LCF-e025-02	5.37	7.22	38.77		0.500	0.80	0.800
LCF-e025-03	5.38	7.23	38.90		0.500	0.80	0.800
LCF-e010-01	5.90	7.35	43.37		0.200	0.80	2.000
LCF-e010-02	5.93	7.29	43.23		0.200	0.80	2.000
LCF-e010-03	5.23	7.35	38.44		0.200	0.80	2.000
LCF-150-01	5.22	7.38	38.52	0	1.500	0.80	0.267
LCF-100-01	5.24	7.47	39.14		1.000	0.80	0.400
LCF-100-02	5.20	7.79	40.51		1.000	0.80	0.400
LCF-100-03	5.65	7.41	41.87		1.000	0.80	0.400
LCF-075-01	4.99	7.41	36.98		0.750	0.80	0.533
LCF-075-02	5.08	7.44	37.80		0.750	0.80	0.533
LCF-075-03	5.84	7.35	42.92		0.750	0.80	0.533
LCF-050-01	5.76	7.41	42.68		0.500	0.80	0.800
LCF-050-02	5.15	7.44	38.32		0.500	0.80	0.800
LCF-050-03	4.99	7.36	36.73		0.500	0.80	0.800
LCF-030-01	5.02	7.41	37.20		0.300	0.80	1.333
LCF-030-02	5.47	7.50	41.03		0.300	0.80	1.333
LCF-030-03	5.37	7.51	40.33		0.300	0.80	1.333
LCF-025-01	5.45	7.40	40.33		0.250	0.80	1.600
LCF-025-02	5.43	7.38	40.07		0.250	0.80	1.600
LCF-025-03	5.63	7.35	41.38		0.250	0.80	1.600
LCF-020-01	5.16	7.48	38.60		0.200	0.80	5.000
LCF-020-02	5.47	7.49	40.97		0.200	0.80	5.000
LCF-020-03	4.49	7.50	33.68		0.200	0.80	5.000
LCF-020-04	6.02	7.34	44.19		0.200	0.80	5.000
LCF-015-01	4.70	7.46	35.06		0.150	0.80	5.000

Table 5.20 Results of fatigue tests performed under strain-control conditions ($R\varepsilon = -1$) for the material from the Eiffel bridge

Specimens	$\Delta\varepsilon$ %	$\Delta\varepsilon^P$ %	$\Delta\varepsilon^E$ %	$\Delta\sigma$MPa	σ_{max}MPa	N_f cycles	$2N_f$ reversals
LCF-viana-050-01	0.500	0.141	0.359	660.37	318.28	2845	5690
LCF-viana-050-02	0.500	0.143	0.357	691.09	344.61	922	1844
LCF-viana-050-03	0.500	0.140	0.360	690.36	338.89	537	1074
LCF-viana-040-01	0.400	0.059	0.341	656.79	324.20	2684	5368
LCF-viana-040-02	0.400	0.073	0.327	632.79	303.50	581	1162
LCF-viana-040-03	0.400	0.065	0.335	637.38	309.97	5697	11394
LCF-viana-030-01	0.300	0.017	0.283	562.65	270.79	80281	160562
LCF-viana-030-02	0.300	0.014	0.286	578.36	293.41	49460	98920
LCF-viana-020-01	0.200	0.002	0.198	408.38	230.90	854096	1708192
LCF-viana-020-02	0.200	0.003	0.197	399.58	269.69	1000000*	2000000*
LCF-viana-075-01	0.750	0.374	0.376	700.35	346.18	164	328
LCF-viana-075-02	0.750	0.362	0.388	726.31	358.49	153	306
LCF_050_01	0.500	0.160	0.340	646.98	309.42	3012	6024
LCF_040_01	0.400	0.060	0.340	639.10	321.99	4154	8308
LCF_030_01	0.300	0.010	0.290	535.88	255.78	80120	160240
LCF_030_02	0.300	0.019	0.281	548.05	275.06	20010	40020
LCF_025_01	0.250	0.007	0.243	463.72	231.61	68911	137822
LCF_007_01	0.256	0.006	0.249	466.70	233.05	68502	137004
LCF_006_01	0.195	0.002	0.194	385.58	193.22	192794	385588
LCF_005_01	0.166	0.00002	0.166	322.59	161.13	5000000*	10000000*
LCF_0055_01	0.188	0.001	0.187	367.51	183.72	984799	1969598
LCF_006_02	0.203	0.00037	0.202	417.24	208.48	684114	1368228
LCF_0055_02	0.188	0.002	0.186	369.31	184.70	914896	1829792
LCF_060_01	0.600	0.236	0.364	700.21	340.76	407	814
LCF_070_01	0.700	0.335	0.365	649.86	305.55	759	1518
LCF_080_01	0.800	0.432	0.368	661.91	316.92	145	290
LCF_100_01	1.000	0.623	0.377	716.16	333.56	50	100

*Run-out

where K' and n' are the cyclic strain hardening coefficient and exponent, respectively; $\Delta\varepsilon^P$ is plastic strain range; and $\Delta\sigma$ is the stress range. In metals, the cyclic strain hardening exponent, n', varies usually in range of 0.05 to 0.25.

Since for some materials and certain strain rates, the stabilized behaviour is not achieved during the cyclic loading, the stabilized hysteresis loops were extracted using the half-life criterion. Therefore, the stabilized behaviour may correspond, for some cases to a pseudo-stabilized behaviour.

Table 5.21 Results of fatigue tests performed under strain-control conditions (Rε = −1) for the material from the Trezói bridge [17]

Specimens	Δε %	$\Delta\varepsilon^P$ %	$\Delta\varepsilon^E$ %	Δσ MPa	σ_{max} MPa	N_f cycles	$2N_f$ reversals
LCF1	1.000	0.694	0.306	574.60	287.20	716	1432
LCF2	0.750	0.470	0.280	555.80	277.47	1626	3252
LCF3	0.500	0.251	0.249	485.80	244.46	5749	11498
LCF4	0.300	0.091	0.209	438.20	219.66	20596	41192
LCF5	0.200	0.001	0.199	436.80	167.79	1031126*	2062252*
LCF6	0.300	0.108	0.192	412.00	204.22	15852	31704
LCF7	0.500	0.241	0.259	542.80	271.93	2586	5172
LCF8	1.500	1.141	0.359	667.40	331.89	414	828
LCF9	2.000	1.623	0.377	715.40	356.16	153	306
LCF10	2.500	2.080	0.420	733.00	369.00	199	398

*Run-out

Table 5.22 Results of fatigue tests performed under strain-control conditions (Rε = −1) for the material from the Luiz I bridge

Specimens	Δε %	$\Delta\varepsilon^P$ %	$\Delta\varepsilon^E$ %	Δσ MPa	σ_{max} MPa	N_f cycles	$2N_f$ reversals
LCF-Luiz1A-050-01	1.000	0.637	0.363	718.98	356.34	17	34
LCF-Luiz1B-050-02	1.000	0.635	0.365	726.64	356.15	28	56
LCF-Luiz1C-0375-01	0.750	0.386	0.364	695.47	340.85	55	110
LCF-Luiz1D-0375-02	0.750	0.367	0.383	720.63	350.28	94	188
LCF-Luiz1E-025-01	0.500	0.161	0.339	675.60	332.83	325	650
LCF-Luiz1F-025-02	0.500	0.165	0.335	662.99	328.87	465	930
LCF-Luiz1G-020-01	0.400	0.077	0.323	659.87	322.89	888	1776
LCF-Luiz1H-020-02	0.400	0.067	0.333	660.51	342.82	987	1974
LCF-Luiz1I-015-01	0.300	0.012	0.288	583.58	288.03	4950	9900
LCF-Luiz1J-015-02	0.300	0.021	0.279	562.44	288.60	97775	195550
LCF-Luiz1K-015-03	0.300	0.023	0.277	548.43	284.42	8551	17102
LCF-Luiz1L-015-04	0.300	0.013	0.287	575.90	279.77	19332	38664
LCF-Luiz1M-010-01	0.200	0.00029	0.19971	411.24	201.86	788684	1577368
LCF-Luiz1N-010-02	0.200	0.00147	0.19853	414.69	201.52	191591	383182
LCF-Luiz1O-010-03	0.200	0.00128	0.19872	415.58	174.75	567651	1135302

Alternatively, the Ramberg–Osgood mathematical representation of the cyclic curve may be presented in the following form, using the total strain amplitude instead of the plastic strain [23]:

$$\frac{\Delta\varepsilon}{2} = \frac{\Delta\varepsilon^E}{2} + \frac{\Delta\varepsilon^P}{2} = \frac{\Delta\sigma}{2E} + \left(\frac{\Delta\sigma}{2K'}\right)^{1/n'} \tag{5.3}$$

Table 5.23 Results of fatigue tests performed under strain-control conditions (Rε = −1 and Rε = 0) for the material from the Fão bridge [13]

Specimens	R_ε	$\Delta\varepsilon$ %	$\Delta\varepsilon^P$ %	$\Delta\varepsilon^E$ %	$\Delta\sigma$ MPa	σ_{max} MPa	N_f cycles	$2N_f$ reversals
LCF-e020-01	−1	0.400	0.101	0.299	561.79	277.43	4416	8832
LCF-e020-02		0.400	0.091	0.309	591.80	287.77	5433	10866
LCF-e020-03		0.400	0.077	0.323	736.84	356.74	8420	16840
LCF-e015-01		0.300	0.027	0.273	507.97	225.52	49300	98600
LCF-e015-02		0.300	0.027	0.273	522.47	230.89	16788	33576
LCF-e015-03		0.300	0.027	0.273	550.33	259.05	27731	55462
LCF-e040-01		0.800	0.441	0.359	657.58	315.05	374	748
LCF-e040-02		0.800	0.355	0.445	730.47	347.96	452	904
LCF_e025-01		0.500	0.192	0.308	631.26	305.33	290	580
LCF-e025-02		0.500	0.171	0.329	623.48	303.19	2108	4216
LCF-e025-03		0.500	0.197	0.303	617.30	297.89	2254	4508
LCF-e010-01		0.200	0.006	0.194	381.34	166.97	530604	1061208
LCF-e010-02		0.200	0.005	0.195	391.23	134.27	663962	1327924
LCF-e010-03		0.200	0.009	0.191	375.88	155.97	74549	149098
LCF-150-01	0	1.500	1.079	0.421	616.59	372.47	143	286
LCF-100-01		1.000	0.554	0.446	824.48	406.74	92	184
LCF-100-02		1.000	0.612	0.388	616.37	305.24	955	1910
LCF-100-03		1.000	0.620	0.380	654.92	324.96	528	1056
LCF-075-01		0.750	0.319	0.431	755.61	361.40	384	768
LCF-075-02		0.750	0.334	0.416	779.19	385.97	250	500
LCF-075-03		0.750	0.373	0.377	699.42	345.92	507	1014
LCF-050-01		0.500	0.169	0.331	572.95	282.58	1106	2212
LCF-050-02		0.500	0.121	0.379	715.77	351.51	3454	6908
LCF-050-03		0.500	0.158	0.342	636.91	331.55	1195	2390
LCF-030-01		0.300	0.074	0.226	531.51	300.39	25560	51120
LCF-030-02		0.300	0.026	0.274	526.50	278.63	22618	45236
LCF-030-03		0.300	0.025	0.275	494.96	272.04	15517	31034
LCF-025-01		0.250	0.069	0.181	467.02	262.92	23878	47756
LCF-025-02		0.250	0.061	0.189	464.76	202.33	76677	153354
LCF-025-03		0.250	0.007	0.243	483.21	237.01	39998	79996
LCF-020-01		0.200	0.050	0.150	372.07	217.32	692177	1384354
LCF-020-02		0.200	0.016	0.184	376.46	214.90	357127	714254
LCF-020-03		0.200	0.019	0.181	385.79	213.07	54116	108232
LCF-020-04		0.200	0.025	0.175	398.56	247.24	306129	612258
LCF-015-01		0.150	0.004	0.146	296.46	158.81	1000000*	2000000*

*Run-out

Fig. 5.10 Some smooth specimens of material from Eiffel bridge used in fatigue tests

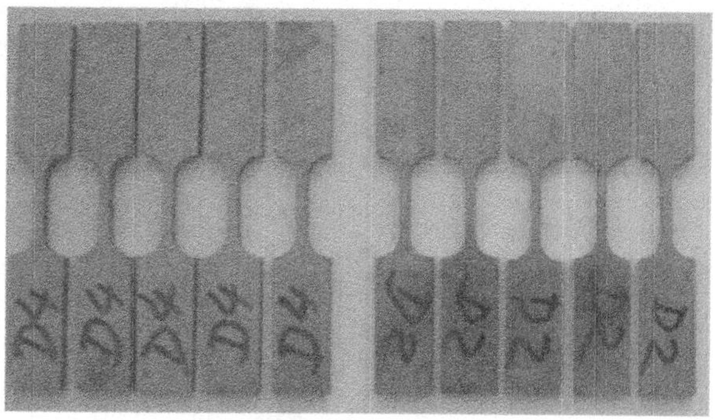

Fig. 5.11 Some smooth specimens of material from Fão bridge used in fatigue tests [13]

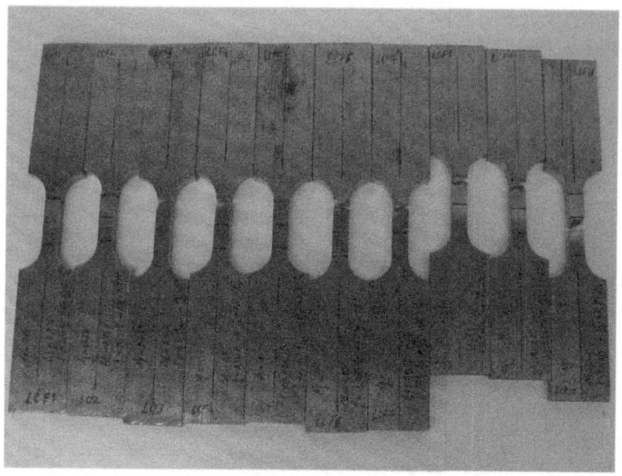

Fig. 5.12 Some smooth specimens of material from Trezói bridge used in fatigue tests

where E is the Young modulus; $\Delta\sigma$ and $\Delta\varepsilon$ are, respectively, the stress and strain ranges; $\Delta\varepsilon^{E}$ and $\Delta\varepsilon^{P}$ are, respectively, the elastic and plastic strain ranges and K' and n' are the same constants of Eq. 5.2.

5.4.1.1 Eiffel Bridge

Figure 5.13 presents the stabilized hysteresis loops together with the cyclic for the sample curve of the material from the Eiffel bridge, obtained for a strain ratio equal to −1 while Fig. 5.14 shows, for the sample material, the superposition of several stabilized hysteresis loops, with the lower tips coinciding at the origin of the graph. Besides the cyclic curve, a cyclic curve scaled by a factor of two is plotted in the figure. The analysis of the figure shows some deviation from the Masing behaviour; however, that deviation may be partly attributed to the scatter in material behaviour due to the high level of heterogeneities found in the microstructures of the material.

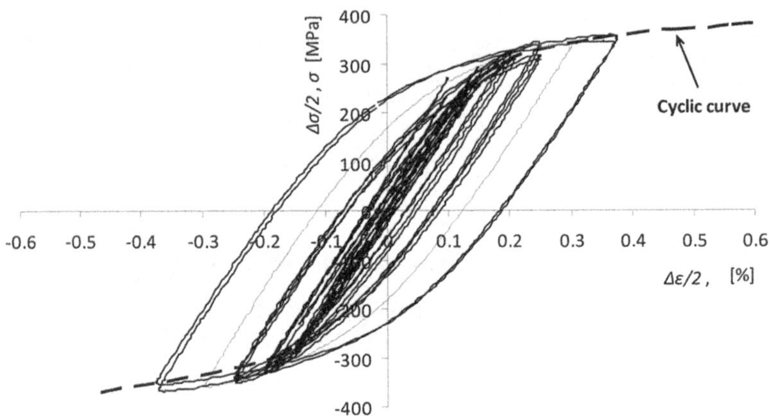

Fig. 5.13 Stabilized hysteresis loops together with the cyclic curve of the Eiffel bridge material, $R_\varepsilon = -1$

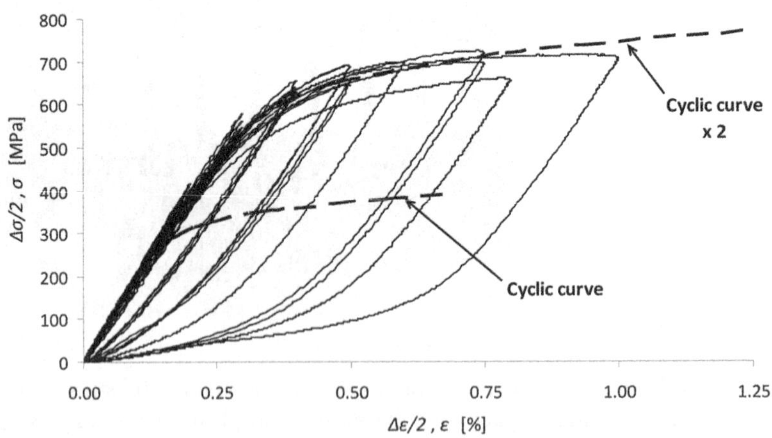

Fig. 5.14 Stabilized hysteresis loops together with the cyclic curve of the Eiffel bridge material, with the lower tips coinciding at the origin, $R_\varepsilon = -1$

Taking into account the scatter effects one may conclude that this material fairly follows the Masing behaviour [18].

Figure 5.15 illustrates the evolution of the cyclic stress amplitude with the number of cycles for the tested strain ranges. The analysis of the data reveals that the material from the Eiffel bridge shows some cyclic hardening for higher strain ranges. This hardening reduces progressively with diminishing applied strain range until it stabilizes when elastic behaviour is achieved.

Figure 5.16 shows the cyclic curve of the material in the form of the stress amplitude versus plastic strain amplitude (Eq. 5.2.). From the analysis of Fig. 5.16 results directly the cyclic strain hardening coefficient (K') and exponent (n') of the material.

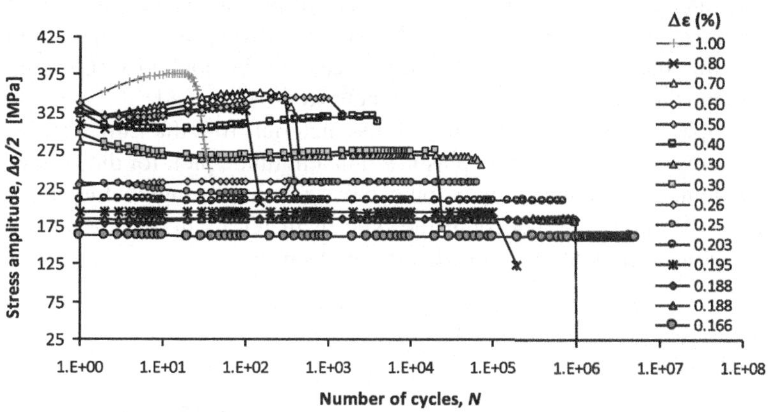

Fig. 5.15 Stress amplitude versus number of cycles for fully reversed strain-controlled tests, $R_\varepsilon =$ -1, of the Eiffel bridge material

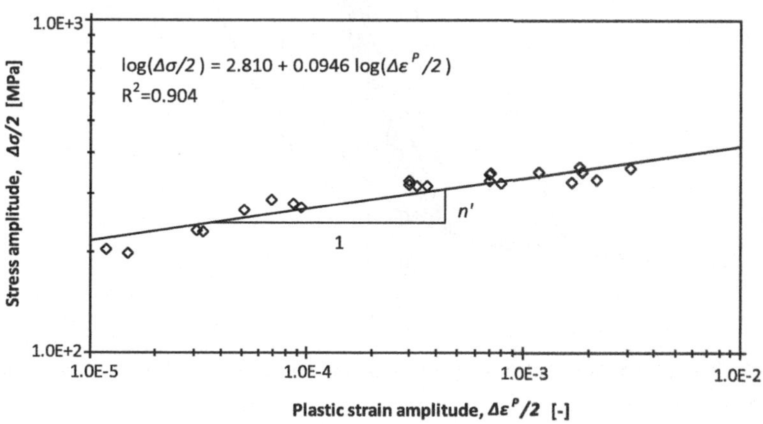

Fig. 5.16 Cyclic curve, $\Delta\sigma/2$ *versus* $\Delta\varepsilon^P/2$ ($R_\varepsilon = -1$), for the Eiffel bridge material

5.4.1.2 Luiz I Bridge

Figure 5.17 presents the stabilized hysteresis loops together with the cyclic curve for the sample of the material from the Luiz I bridge, obtained for a strain ratio equal to −1 [18], while Fig. 5.18 shows, for the same material, the superposition of several stabilized hysteresis loops, with the lower tips coinciding at the origin of the graph. This material from the Luiz I bridge shows some deviation from the Masing behaviour; this deviation may be also partially attributed to the scatter in material behaviour. Therefore, the material from the Luiz I bridge shows a cyclic behaviour that fairly follows the Masing behaviour, similarly to the material of the Eiffel bridge. Figure 5.19 illustrates the evolution of the cyclic stress amplitude with the number of cycles for the tested strain ranges. The analysis of the data reveals that the material from the Luiz I bridge shows some slight cyclic hardening for higher strain ranges. However, this hardening fades with the decrease in the applied strain range, until stabilization is achieved, when the elastic behaviour is attained.

Figure 5.20 shows the cyclic curve of the material from the Luiz I bridge in the form of a stress amplitude versus plastic strain amplitude plot, for direct assessment of the hardening constants, namely the cyclic strain hardening coefficient (K') and exponent (n'). Taking into account the type of materials under investigation, the determination coefficient is assumed very satisfactory.

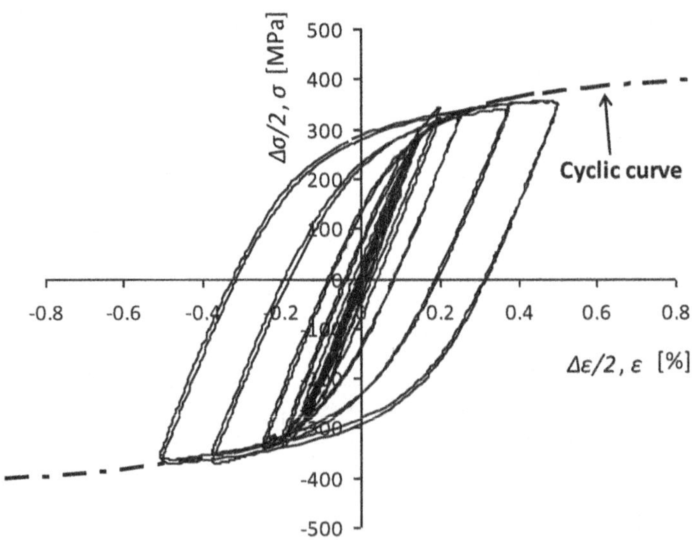

Fig. 5.17 Stabilized hysteresis loops together with the cyclic curve of the Luiz I bridge material, $R_\varepsilon = -1$

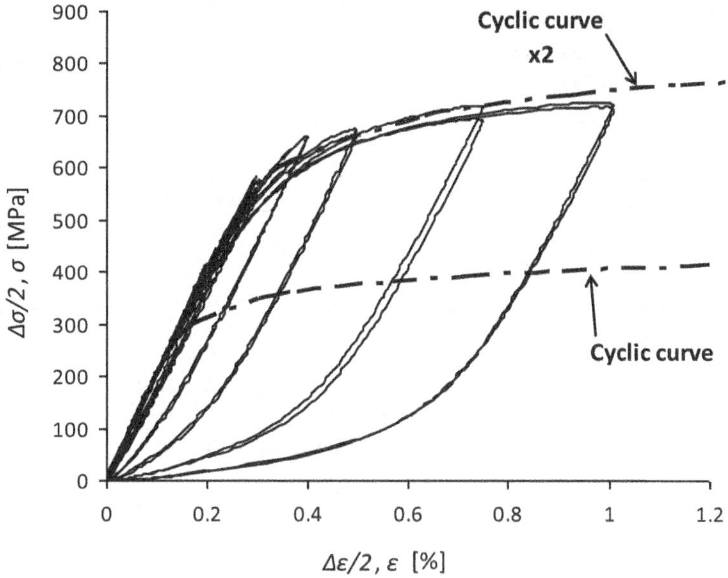

Fig. 5.18 Stabilized hysteresis loops together with the cyclic curve of the Luiz I bridge material, with the lower tips coinciding at the origin, $R_\varepsilon = -1$

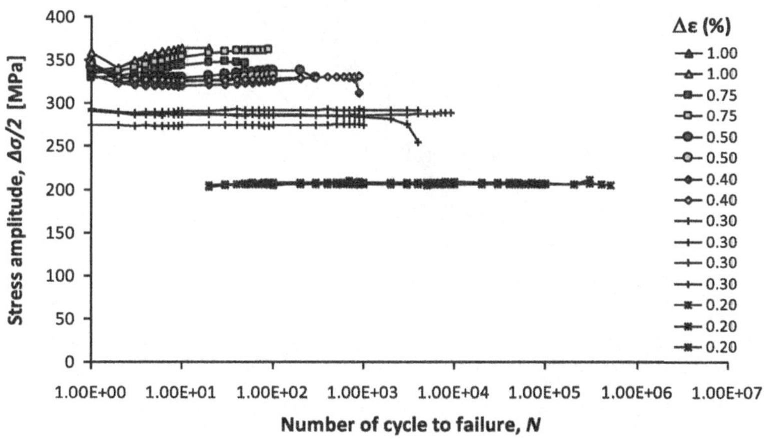

Fig. 5.19 Stress amplitude versus number of cycles for fully reversed strain-controlled tests, $R_\varepsilon = -1$, of the Luiz I bridge material

5.4.1.3 Fão Bridge

Figure 5.21 presents the stabilized hysteresis loops together with the cyclic curve for the material sample from the Fão bridge obtained for strain ratios equal to −1 [13, 18], while Figs. 5.22 and 5.23 show for the same material, the superposition of

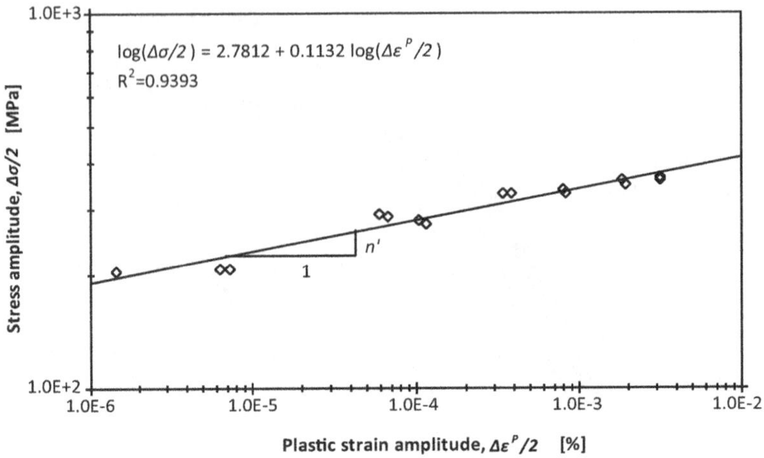

Fig. 5.20 Cyclic curve, $\Delta\sigma/2$ *versus* $\Delta\varepsilon^P/2$ ($R_\varepsilon = -1$), for the Luiz I bridge material

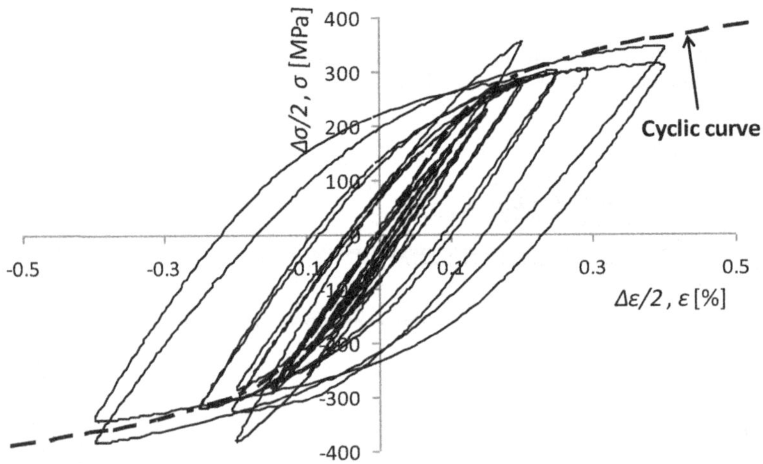

Fig. 5.21 Stabilized hysteresis loops together with the cyclic curve of the Fão bridge material, $R_\varepsilon = -1$

several stabilized hysteresis loops, with the lower tips coinciding at the origin of the graph, obtained for strain ratios equal to −1 and 0, respectively. The significant scatter observed in the hysteresis loops hinders the assessment of the Masing behaviour of the material. This scatter is noticeable in the material from the Fão bridge showing two anomalous hysteresis loops, for strain ratio equal −1. Neglecting these two hysteresis loops the material shows Masing behaviour for strain ratio equal −1. For null strain ratio, the material from Fão bridge shows a clear non-Masing behaviour.

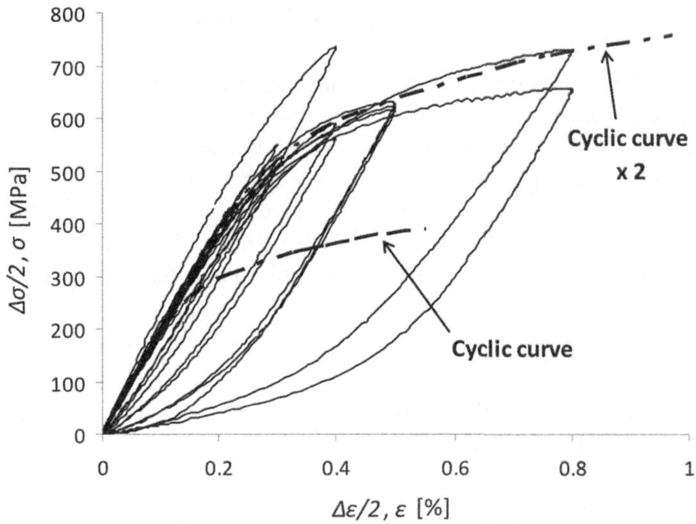

Fig. 5.22 Stabilized hysteresis loops together with the cyclic curve of the Fão bridge material, with the lower tips coinciding at the origin, $R_{\varepsilon} = -1$

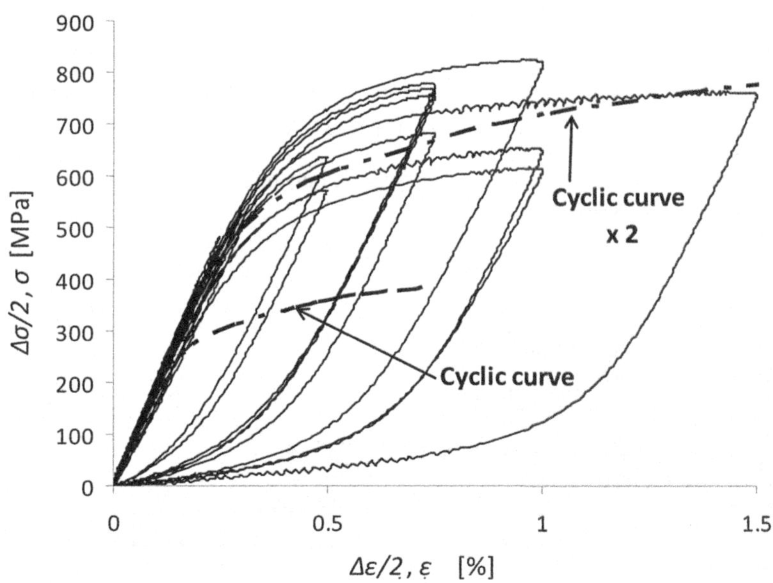

Fig. 5.23 Stabilized hysteresis loops together with the cyclic curve of the Fão bridge material, with the lower tips coinciding at the origin, $R_{\varepsilon} = 0$

It seems that hysteresis loops of the material from Fão bridge are sensitive to the strain.

Figures 5.24 and 5.25 illustrate the evolution of the cyclic stress amplitude with the number of cycles for the tested strain ranges, for strain ratios equal to −1 and 0, respectively. The analysis of the data reveals that material from the Fão bridge displays cyclic hardening for higher strain ranges, but this cyclic hardening reduces progressively with decreasing applied strain range until it stabilizes for fully elastic behaviour. This behaviour is similar to the behaviour observed for the materials from the Eiffel and Luiz I bridges. It is interesting to note that the material from the

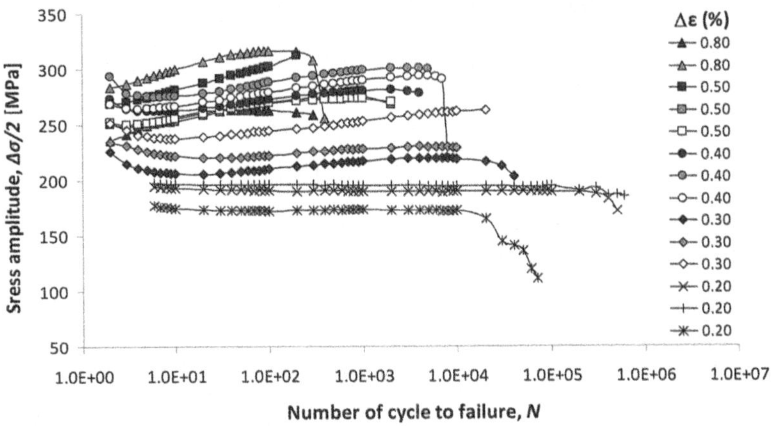

Fig. 5.24 Stress amplitude versus number of cycles for fully reversed strain-controlled tests, $R_\varepsilon = -1$, of the Fão bridge material

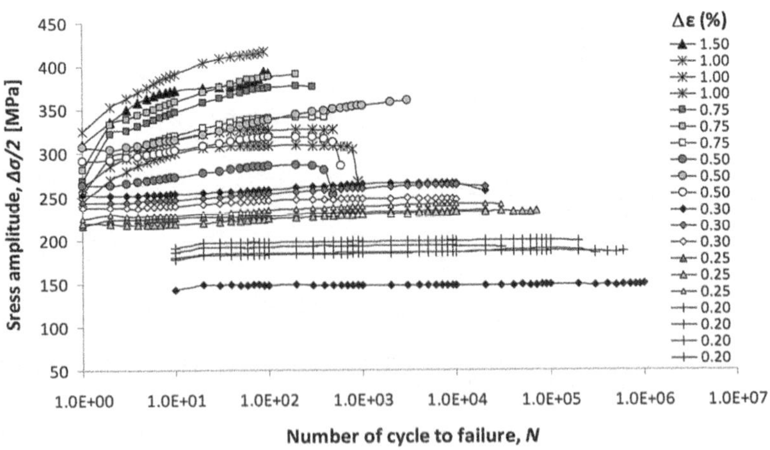

Fig. 5.25 Stress amplitude versus number of cycles for alternate strain-controlled tests, $R_\varepsilon = 0$, of the Fão bridge material

Fão bridge shows higher hardening for null strain ratios than for fully-reverse strain loading conditions.

Figures 5.26, 5.27 and Fig. 5.28 show the cyclic curves of the material from the Fão bridge. The stress amplitude is plotted against the plastic strain amplitude, for $R_\varepsilon = -1$, $R_\varepsilon = 0$ and $R_\varepsilon = -1 + R_\varepsilon = 0$, respectively. From the analysis of these figures results directly the cyclic strain hardening constants of the material from the Fão bridge. Comparing with the previous materials, lower determination coefficients in the cyclic curve assessment were observed for this material, which is consistent with the higher scatter observed in hysteresis loops depicted in the Figs. 5.22 and 5.23.

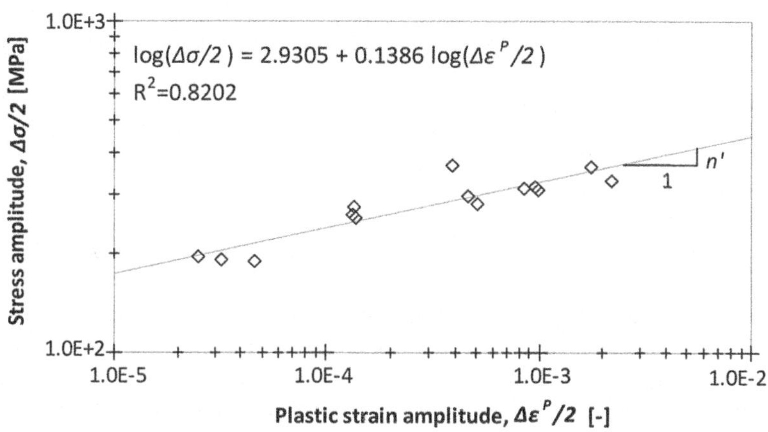

Fig. 5.26 Cyclic curve, $\Delta\sigma/2$ versus $\Delta\varepsilon^P/2$ ($R_\varepsilon = -1$), for the Fão bridge material

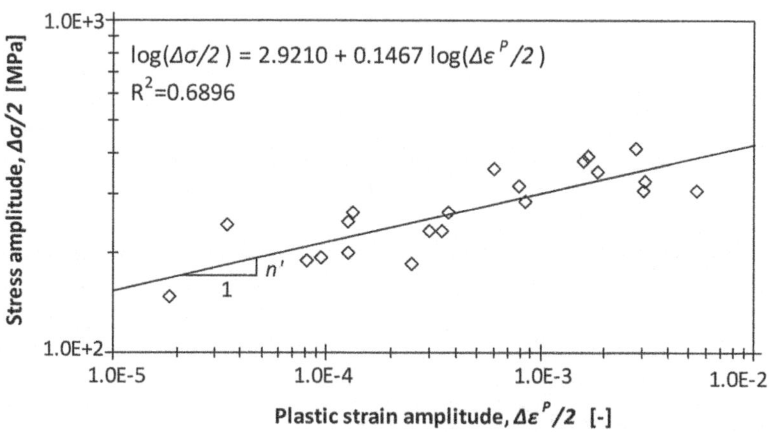

Fig. 5.27 Cyclic curve, $\Delta\sigma/2$ versus $\Delta\varepsilon^P/2$ ($R_\varepsilon = 0$), for the Fão bridge material

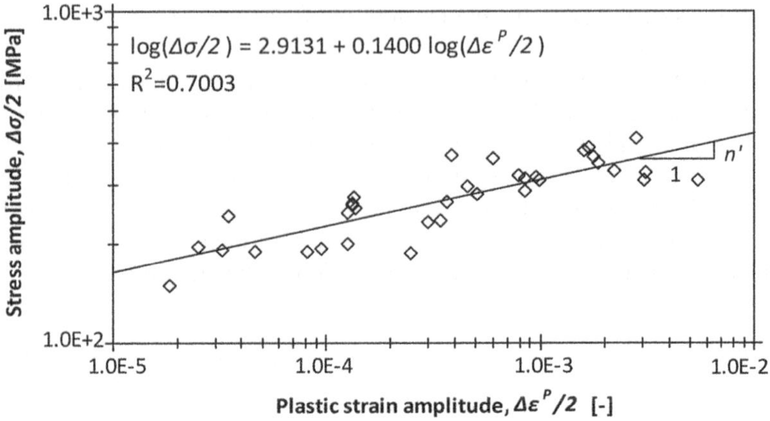

Fig. 5.28 Cyclic curve, $\Delta\sigma/2$ versus $\Delta\varepsilon^P/2$ ($R_\varepsilon = -1 + R_\varepsilon = 0$), for the Fão bridge material

5.4.1.4 Trezói Bridge

Figure 5.29 presents the stabilized hysteresis loops superimposed with the cyclic curve of the material from the Trezoi bridge [17, 18], obtained for a strain ratio equal to -1. Figure 5.30 shows the superposition of several stabilized hysteresis loops with the lower tips coinciding at the origin of the graph of the material, together with the cyclic curve magnified by a factor of two. The analysis of this figure shows clearly that the material from the Trezói bridge does not follow the Masing behaviour.

Figure 5.31 illustrates the evolution of the cyclic stress amplitude with the number of cycles for the tested strain ranges and for a strain ratio equal to −1. The material from the Trezói bridge shows a clear cyclic softening for strain ranges below 1.5%.

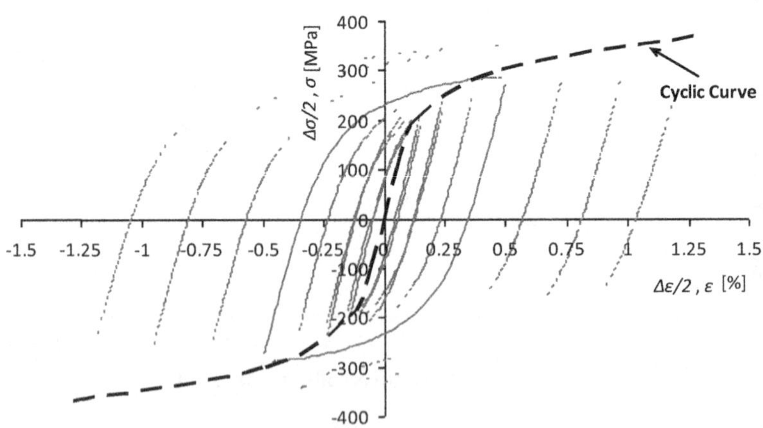

Fig. 5.29 Stabilized hysteresis loops together with the cyclic curve of the Trezói bridge material, $R_\varepsilon = -1$

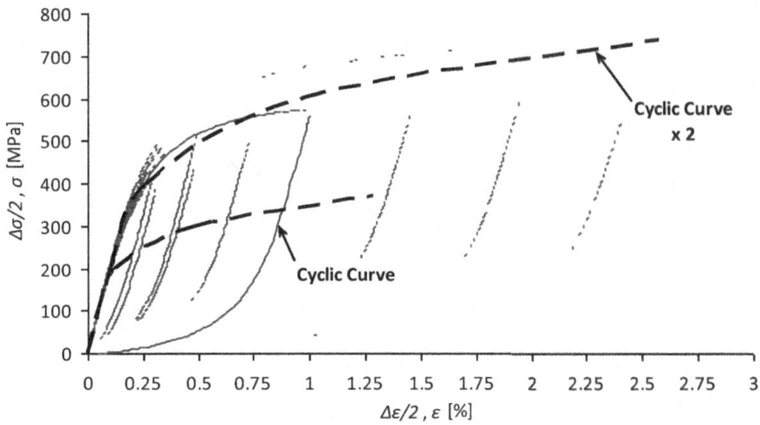

Fig. 5.30 Stabilized hysteresis loops together with the cyclic curve of the Trezói bridge material, with the lower tips coinciding at the origin, $R_{\varepsilon} = -1$

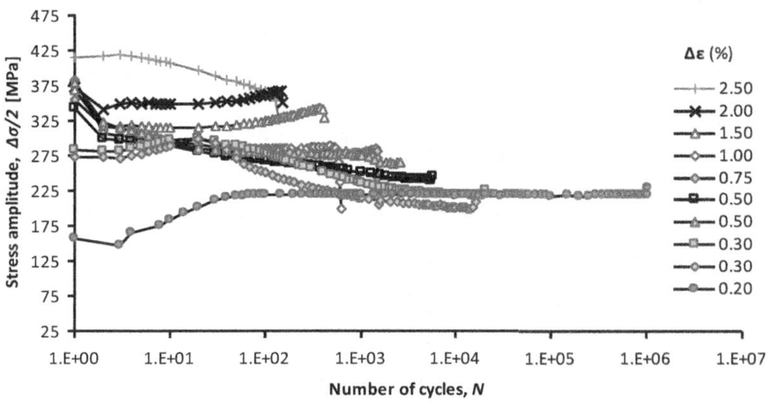

Fig. 5.31 Stress amplitude versus number of cycles for fully reversed strain-controlled tests, $R_{\varepsilon} = -1$, of the Trezói bridge material

For strain ranges above 1.5%, the cyclic behaviour is not clearly defined since for a strain range of 2.5% a cyclic softening is observed whereas for strain ranges of 2 and 1.5%, a cyclic hardening is verified, after some stabilization. If a comparison between all materials is made only for strain ranges below 1%, but involving plasticity, it is clear that the material from the Trezói bridge shows a cyclic softening while the other materials show cyclic hardening. The material from the Trezói bridge shows higher cyclic hardening values.

Figure 5.32 shows the cyclic curve of the material from the Trezói, bridge for a strain ratio equal to −1. From the analysis of Fig. 5.32 results directly the cyclic strain hardening coefficient and exponent. Figure 5.33 compares the cyclic curves of the four materials from the Eiffel, Luiz I, Fão and Trezói bridges. The material

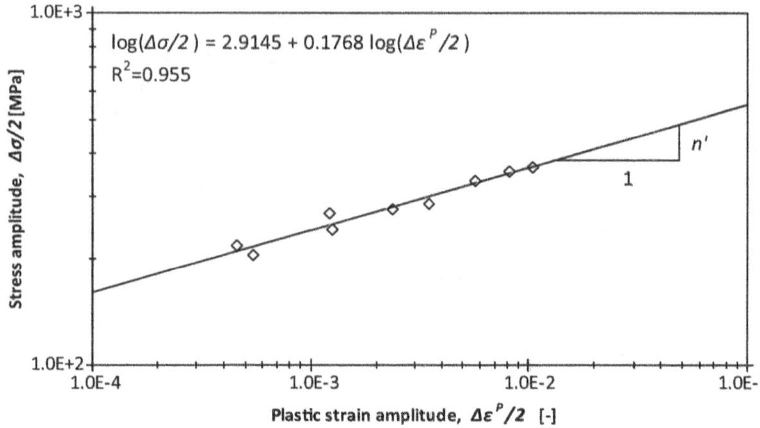

Fig. 5.32 Cyclic curve, $\Delta\sigma/2$ versus $\Delta\varepsilon^P/2$ ($R_\varepsilon = -1$), for the Trezói bridge material

Fig. 5.33 Comparison of the cyclic curves for the materials studied

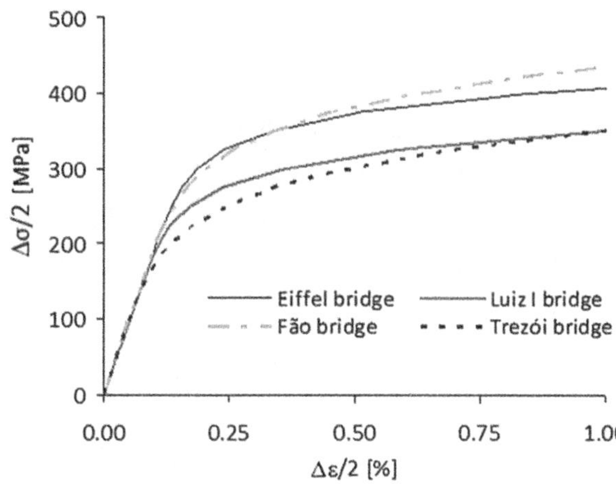

from the Trezói bridge shows the lowest cyclic yield stress, which means that this material experiences a significant cyclic softening with respect to the others. The other materials show higher cyclic yield stresses. The materials from Eiffel and Fão bridge show similar cyclic behaviours, mainly for lower strain amplitudes. The materials from the Trezói and Luiz I bridges also show similar cyclic behaviours, but in this case for high strain amplitudes.

5.4.2 *Evaluation of Fatigue Behaviour Based on Strain-Life Relations*

Low-cycle fatigue results are very often represented using the relation between the strain amplitude and the number of reversals to failure, 2Nf, usually assumed to correspond to the initiation of a macroscopic crack. The classical way to express the strain-life data is to use the superposition of Coffin [24] and Manson [25] equation with the Basquin [26] equation, usually known as Morrow relation [22]. The Coffin and Manson equation relates the plastic strain amplitude, $\Delta\varepsilon^P/2$, to the number of reversals at crack initiation, $2N_f$:

$$\frac{\Delta\varepsilon^P}{2} = \varepsilon'_f (2N_f)^c \tag{5.4}$$

where ε'_f and c are, respectively, the fatigue ductility coefficient and fatigue ductility exponent. The Coffin-Manson relation is well suited for the low-cycle fatigue domain whereas the Basquin relation [26] that relates the elastic strain amplitude, $\Delta\varepsilon^E/2$, to the number of reversals to failure $2N_f$, is best suited for high cycle fatigue:

$$\frac{\Delta\varepsilon^E}{2} = \frac{\Delta\sigma}{2E} = \frac{\sigma'_f}{E}(2N_f)^b \tag{5.5}$$

where σ'_f is the fatigue strength coefficient, b is the fatigue strength exponent and E is the Young modulus. The number of reversals corresponding to the transition between low and high cycle fatigue regimes is characterized by total strain amplitude composed by equal components of elastic and plastic strain amplitudes. Lives below this transition value are dictated by ductility properties while lives above this transition value are dictated by strength properties. Morrow [22] suggested the superposition of Eqs. 5.4 and 5.5, resulting in a more general equation, valid for low and high-cycle fatigue regimes:

$$\frac{\Delta}{\varepsilon}2 = \frac{\Delta\varepsilon^E}{2} + \frac{\Delta\varepsilon^P}{2} = \frac{\sigma'_f}{E}(2N_f)^b + \varepsilon'_f (2N_f)^c \tag{5.6}$$

Alternatively to the Morrow relation, the Smith-Watson-Topper fatigue damage parameter [38] can be used, which shows the following form:

$$\sigma_{max}\frac{\Delta\varepsilon}{2} = SWT = \frac{\left(\sigma'_f\right)^2 (2N_f)^{2b}}{E} + \sigma'_f \varepsilon'_f (2N_f)^{b+c} \tag{5.7}$$

where σ_{max} is the maximum stress of the cycle and SWT is the damage parameter. Both Morrow and Smith-Watson-Topper models are deterministic models and are used to represent the average fatigue behaviour of the bridge materials based on the available experimental data.

5.4.2.1 Strain-Life Results and Discussion for Materials from Ancient Portuguese Riveted Steel Bridges

Figures 5.34, 5.35, 5.36, 5.37, 5.38 and 5.39 present the strain-life data for the materials from the Eiffel, Luiz I, Fão and Trezói bridges. The total strain, i.e. elastic strain plus plastic strain, versus life relations is considered. The data are correlated based on

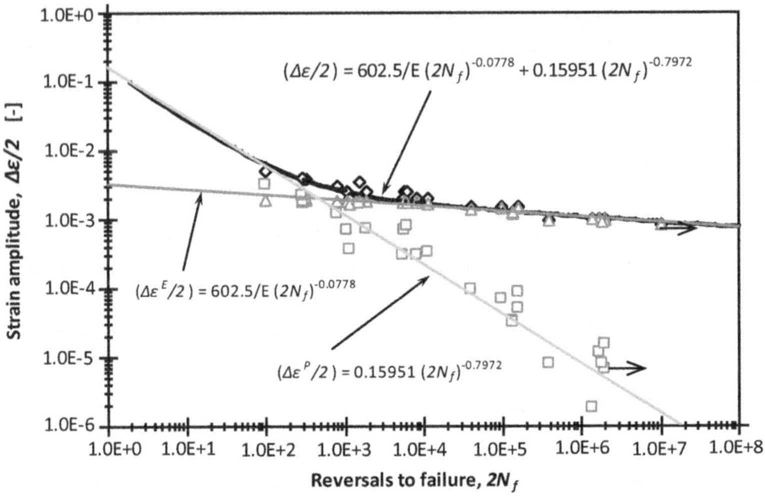

Fig. 5.34 Strain-life curves according to Morrow's model for the material from the Eiffel bridge, $R_\varepsilon = -1$

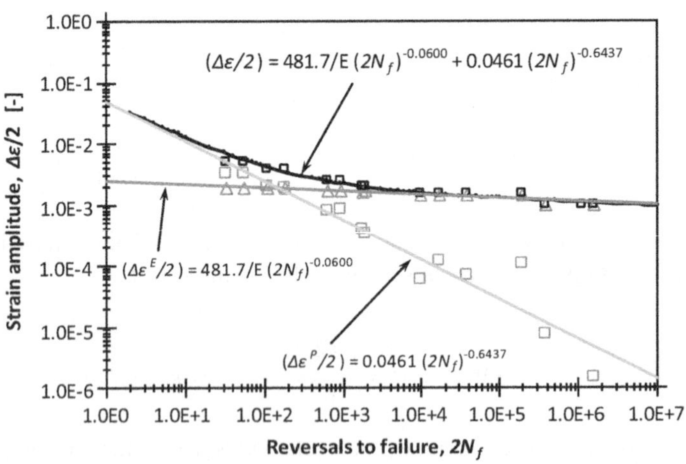

Fig. 5.35 Strain-life curves according to Morrow's model for the material from the Luiz I bridge, $R_\varepsilon = -1$

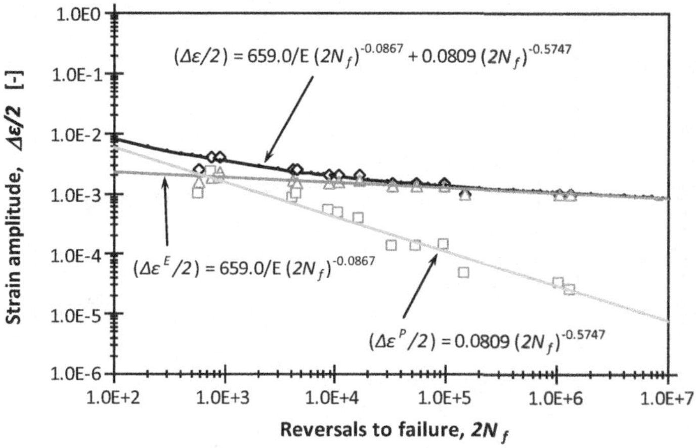

Fig. 5.36 Strain-life curves according to Morrow's model for the material from Fão bridge, $R_\varepsilon = -1$

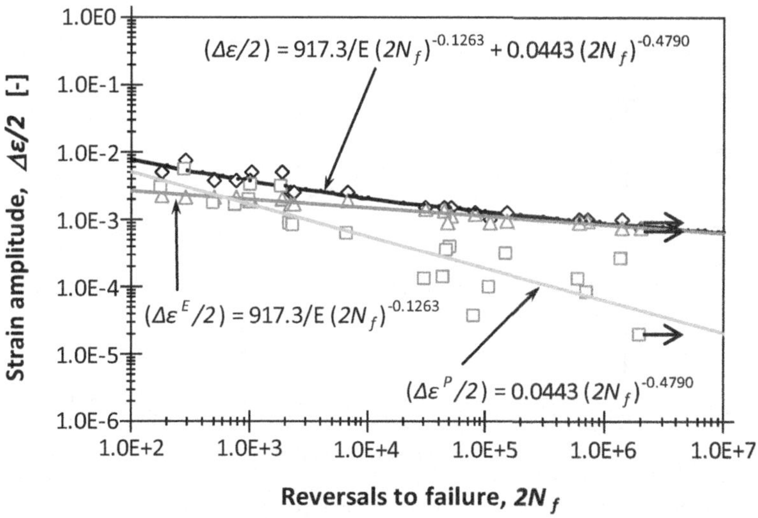

Fig. 5.37 Strain-life curves according to Morrow's model for the material from Fão bridge, $R_\varepsilon = 0$

the Coffin-Manson, Basquin and Morrow models. The materials from the centenary bridges shows very low transition lives ($2N_t$), which are bellow the 1×10^3 reversals. The material from the Trezói bridge exhibits a number of transition reversals one order of magnitude higher possibly because the fatigue ductility coefficient is significantly higher in the material from the Trezói bridge than in the materials from the other centenary bridges. The fatigue ductility coefficient increases for younger materials, according to better techniques achieved in the steel production. Regarding

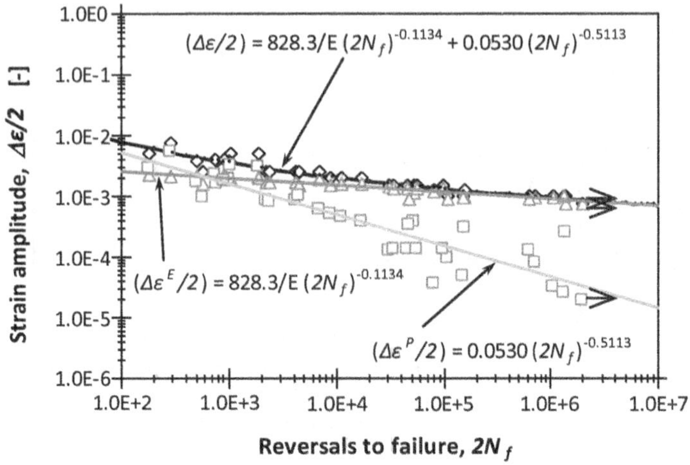

Fig. 5.38 Strain-life curves according to Morrow's model for the material from Fão bridge, $R_\varepsilon = -1 + R_\varepsilon = 0$

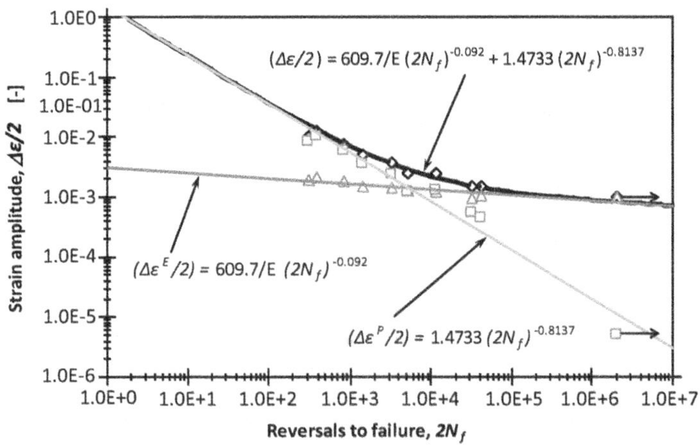

Fig. 5.39 Strain-life curves according to Morrow's model for the material from the Trezói bridge, $R_\varepsilon = -1$

the material from the Fão bridge, higher fatigue ductility was verified for the strain ratio equal to −1. For the material from the Fão bridge, two distinct strain-ratios were tested, namely a null-strain ratio and a fully reversible strain history. Besides the individual treatment of the test data for each strain ratio, Fig. 5.36 presents the stain-life data for the material from the Fão bridge resulting from the conjunction of both strain ratios, $R_\varepsilon = -1 \& R_\varepsilon = 0$ [13].

Tables 5.24, 5.25, 5.26, 5.27, 5.28 and 5.29 summarize the main parameters

Table 5.24 Low-cycle fatigue constants for the material from the Eiffel bridge, $R_\varepsilon = -1$

Fatigue strength coefficient, σ'_f [MPa]	*Fatigue strength exponent, b* [–]	*Determination coefficient, R^2* [–]
602.5	−0.0778	0.8985
Fatigue ductility coefficient, ε'_f [–]	*Fatigue ductility exponent, c* [–]	*Determination coefficient, R^2* [–]
0.1595	−0.7972	0.9002
Transition number of reversals, $2N_f$ [–]	*Transition total strain amplitude, $\Delta\varepsilon_t/2$* [%]	σ'_f/E [–]
475	0.3104	0.00312

Table 5.25 Low-cycle fatigue constants for the material from the Luiz I bridge, $R_\varepsilon = -1$

Fatigue strength coefficient, σ'_f [MPa]	*Fatigue strength exponent, b* [–]	*Determination coefficient, R^2* [–]
481.7	−0.0600	0.8864
Fatigue ductility coefficient, ε'_f [–]	*Fatigue ductility exponent, c* [–]	*Determination coefficient, R^2* [–]
0.0461	−0.6437	0.9157
Transition number of reversals, $2N_f$ [–]	*Transition total strain amplitude, $\Delta\varepsilon_t/2$* [%]	σ'_f/E [–]
150	0.3631	0.00243

Table 5.26 Low-cycle fatigue constants for the material from the Fão bridge, $R_\varepsilon = -1$

Fatigue strength coefficient, σ'_f [MPa]	*Fatigue strength exponent, b* [–]	*Determination coefficient, R^2* [–]
659.0	−0.0867	0.7836
Fatigue ductility coefficient, ε'_f [–]	*Fatigue ductility exponent, c* [–]	*Determination coefficient, R^2* [–]
0.0809	−0.5747	0.9411
Transition number of reversals, $2N_f$ [–]	*Transition total strain amplitude, $\Delta\varepsilon_t/2$* [%]	σ'_f/E [–]
695	0.3762	0.00332

Table 5.27 Low-cycle fatigue constants for the material from the Fão bridge, $R_\varepsilon = 0$

Fatigue strength coefficient, σ'_f [MPa]	*Fatigue strength exponent, b* [–]	*Determination coefficient, R^2* [–]
917.3	−0.1263	0.9226
Fatigue ductility coefficient, ε'_f [–]	*Fatigue ductility exponent, c* [–]	*Determination coefficient, R^2* [–]
0.0443	−0.4790	0.7809
Transition number of reversals, $2N_f$ [–]	*Transition total strain amplitude, $\Delta\varepsilon_t/2$* [%]	σ'_f/E [–]
609	0.4109	0.00462

Table 5.28 Low-cycle fatigue constants for the material from the Fão bridge, $R_\varepsilon = -1 + R_\varepsilon = 0$

Fatigue strength coefficient, σ'_f [MPa]	*Fatigue strength exponent, b* [–]	*Determination coefficient, R^2* [–]
828.3	−0.1134	0.8645
Fatigue ductility coefficient, ε'_f [–]	*Fatigue ductility exponent, c* [–]	*Determination coefficient, R^2* [–]
0.0530	−0.5113	0.8233
Transition number of reversals, $2N_f$ [–]	*Transition total strain amplitude, $\Delta\varepsilon_t/2$* [%]	σ'_f/E [–]
334	0.4874	0.00417

Table 5.29 Low-cycle fatigue constants for the material from the Trezói bridge, $R_\varepsilon = -1$

Fatigue strength coefficient, σ'_f [MPa]	*Fatigue strength exponent, b* [–]	*Determination Coefficient, R^2* [–]
609.7	−0.0920	0.7870
Fatigue ductility coefficient, ε'_f [–]	*Fatigue ductility exponent, c* [–]	*Determination Coefficient, R^2* [–]
1.4733	−0.8137	0.9582
Transition number of reversals, $2N_f$ [–]	*Transition total strain amplitude, $\Delta\varepsilon_t/2$* [%]	σ'_f/E [–]
5184	0.2797	0.00307

derived from the low-cycle fatigue tests, including the constants from the strain-life equation and the cyclic curve of the materials from the Eiffel, Luiz I, Fão and Trezói bridges [13, 17, 18]. The number of reversals corresponding to the transition from an elastic dominant behaviour to a plastic dominant behaviour is also pointed out in the referred tables.

Figures 5.40, 5.41, 5.42, 5.43, 5.44 and 5.45 present the fitting of the fatigue data using the Smith-Watson-Topper damage parameter ($= \sigma_{max} - \varepsilon_a$) resulting in a good correlation with the experimental data. This damage parameter accounts for the mean stress effects, since it depends on the maximum stress of the cycle. The referred figures also include the representation of Eq. 5.7.

Figures 5.46, 5.47 and 5.48 compare the Morrow equation between the materials from the bridges. The analysis of the results confirms that the material from the Trezói bridge shows the highest fatigue resistance for low to high-cycle fatigue regimes. This higher fatigue resistance is mainly due to the higher ductility as illustrates Fig. 5.46. All materials show very similar elastic strain-life relations.

Figure 5.49 shows that total strain-life curves for the material from the Fão bridge are not sensitive to the strain ratio. This is the result of the well know behaviour in cyclic plasticity of metals that corresponds to the cyclic mean stress relaxation occurring for non-symmetric strain local conditions.

Figure 5.50 exhibits typical fracture surfaces from the smooth specimens used to assess the strain-life fatigue data and the cyclic elastoplastic behaviours of the bridge materials [18]. The fracture surfaces of the material from Luiz I bridge shows clearly a lamellar structure corresponding to lined-up inclusions. Fracture surface of the specimen from the material from the Eiffel bridge also shows some lamellar structure. The fatigue crack propagation surfaces are clearly identified in the specimens from

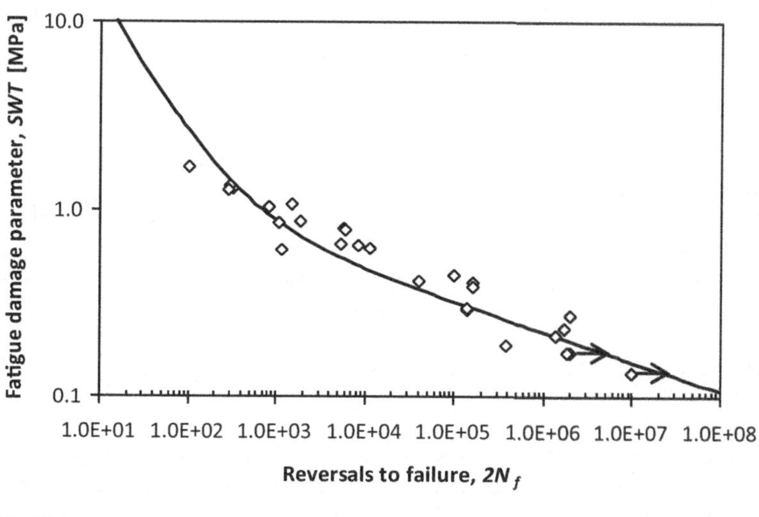

Fig. 5.40 Fitting of the *SWT* model to the lifetime for the material data from Eiffel bridge, $R_\varepsilon = -1$

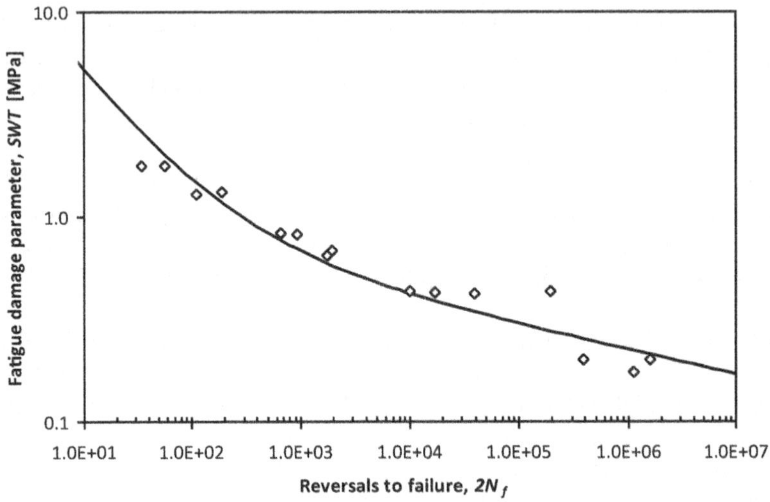

Fig. 5.41 Fitting of the *SWT* model to the lifetime for the material data from Luiz I bridge, $R_\varepsilon = -1$

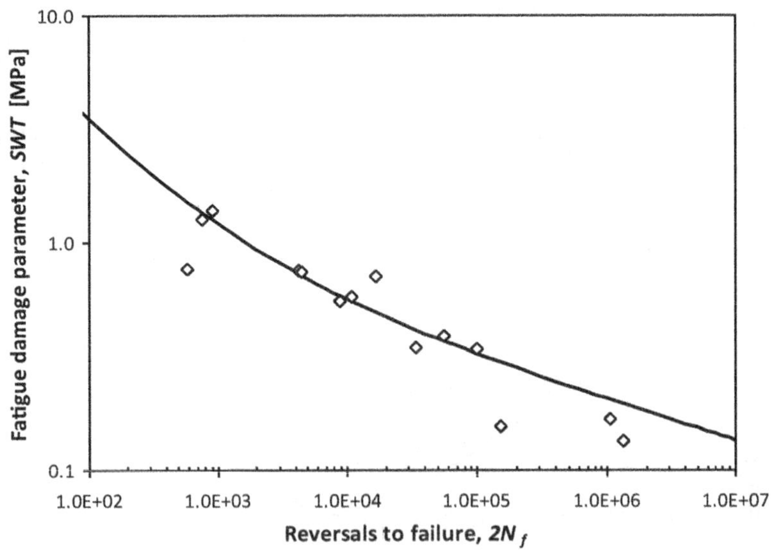

Fig. 5.42 Fitting of the *SWT* model to the lifetime for the material data from Fão bridge, $R_\varepsilon = -1$

the Fão and Trezói bridges. The fracture surface of the specimen made of material from Trezói bridge shows lower levels of roughness, which is consistent with finer microstructure of the material and higher ductility.

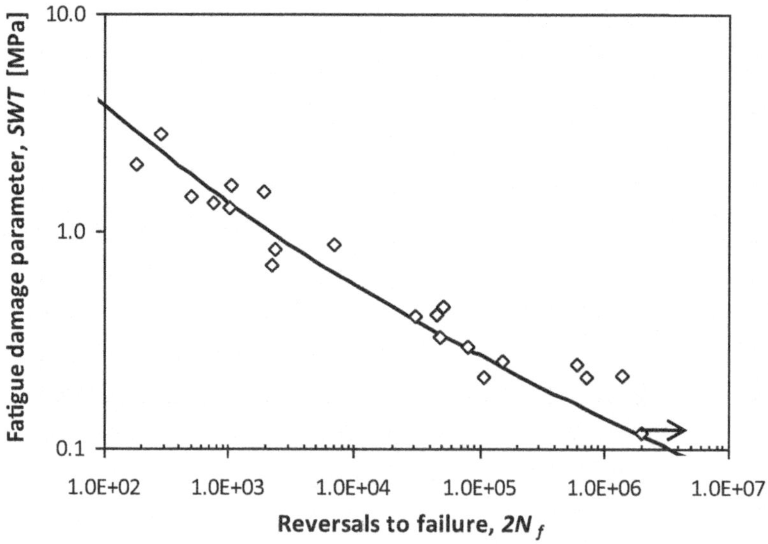

Fig. 5.43 Fitting of the *SWT* model to the lifetime for the material data from Fão bridge, $R_\varepsilon = 0$

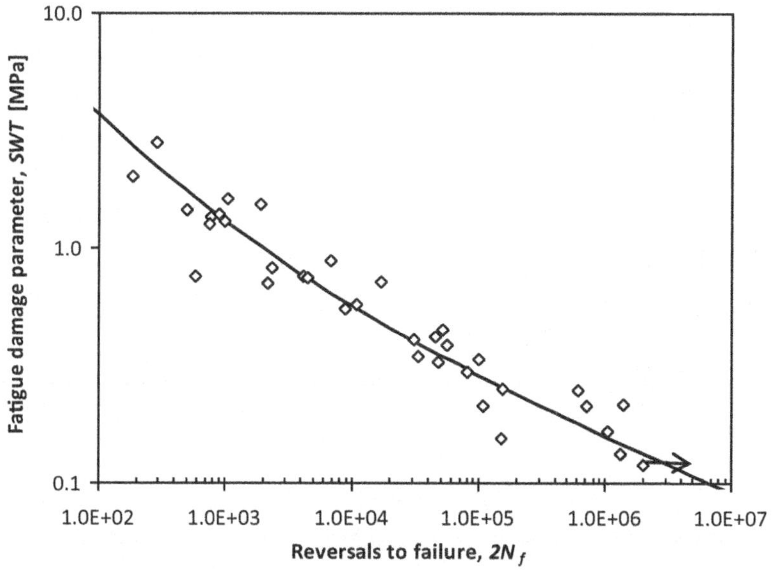

Fig. 5.44 Fitting of the *SWT* model to the lifetime for the material data from the Fão bridge, $R_\varepsilon = -1 + R_\varepsilon = 0$

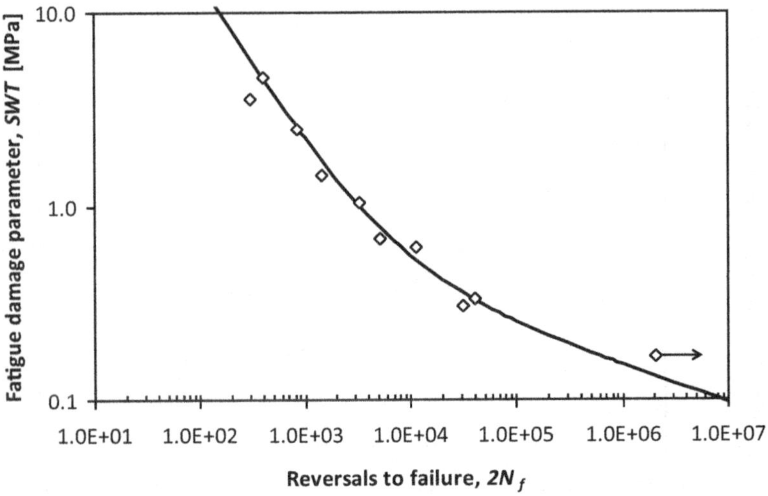

Fig. 5.45 Fitting of the *SWT* model to the lifetime for the material data from the Trezói bridge, $R_\varepsilon = -1$

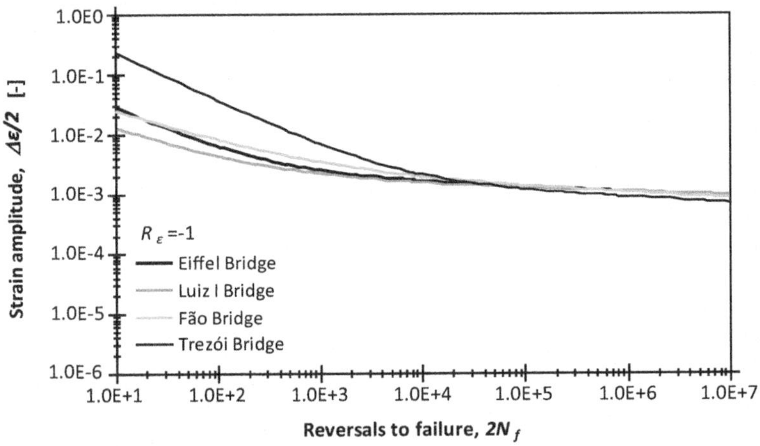

Fig. 5.46 Comparison of the total strain-life using Morrow curves for all materials, $R_\varepsilon = -1$

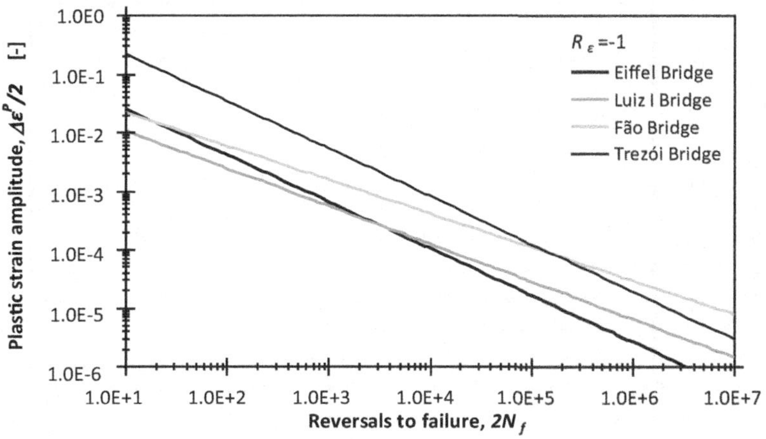

Fig. 5.47 Comparison of the plastic strain-life using Coffin-Manson curves for all materials, $R_{\varepsilon} = -1$

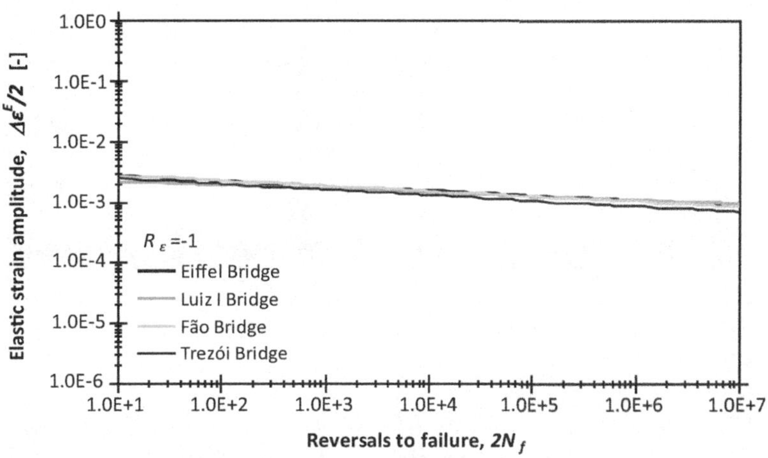

Fig. 5.48 Comparison of the strain-life using Basquin curves for all materials, $R_{\varepsilon} = -1$

5.5 Fatigue Crack Propagation Rates of the Materials

5.5.1 Experimental Details

Fatigue crack growth tests were conducted on old bridge materials according to the ASTM 647-99 standard [27] in order to assess the fatigue crack propagation rates for these materials. An experimental program was carried out using samples of the

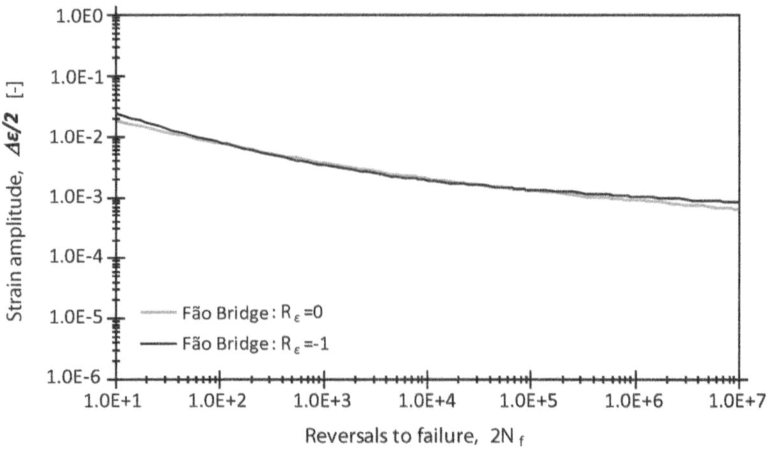

Fig. 5.49 Comparison of the total strain-life using Morrow curves for the material from the Fão bridge

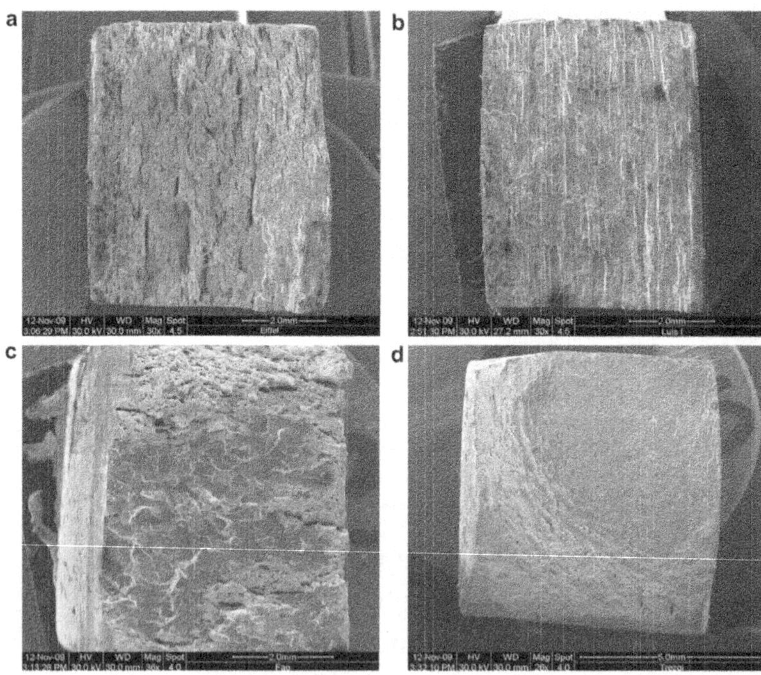

Fig. 5.50 Typical fracture surfaces of the smooth specimens tested under strain control: **a** Eiffel; **b** Luiz I; **c** Fão; **d** Trezói bridges [18]

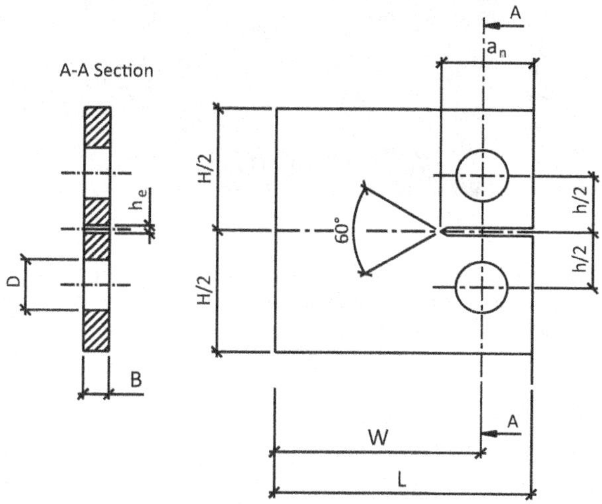

Fig. 5.51 *CT* specimens geometry used in fatigue crack growth tests for materials from Eiffel, Fão, Pinhão and Trezói bridges

materials from the five bridges [11–13, 15, 17, 18]. Compact tension specimens (CT specimens), see Fig. 5.51, were prepared with materials from Eiffel, Fão, Pinhão and Trezói bridges. Middle tension specimens were used for the material from Luiz I bridge (W = 40 mm; t = 10 mm). Due to limitations in material availability, distinct dimensions were adopted for the specimens. Table 5.30 shows the dimensions for the compact tension specimens and the total number of specimens tested for each material. All tests were performed in air, at room temperature, under a sinusoidal waveform with a frequency of 20 Hz for all materials with the exception of the material from Luiz I bridge, which was tested under a frequency of 10 Hz. The crack growth was measured on both faces of the specimens by direct visual inspection using two travelling microscopes with accuracy of 0.001 mm (see Fig. 5.52). The tests were performed in load control conditions in an Instron close-loop servo hydraulic machine, model 8801, rated to 100 KN (see Fig. 5.52).

A total of 42 specimens were tested: five from the Eiffel bridge (one according to the longitudinal direction and four according to the transverse direction); four from the Luiz I bridge; 12 specimens from the Fão bridge; 13 specimens from the Pinhão

Table 5.30 Nominal dimensions (see Fig. 3.51 for reference dimensions) of the compact tension specimens and total number of the specimens used in fatigue tests for each material of the bridges

Bridge material	No. of specimens	*W* mm	*L* mm	*H/2* mm	*h/2* mm	*B* mm	*D* mm	h_e mm	a_n mm
Eiffel	5	40.0	50.0	24.0	11.00	4.35	10.0	1.6	8.0
	12	40.0	50.0	24.0	11.00	5.00	10.0	1.6	8.0
Fão	12	50.0	62.5	30.0	13.75	8.00	12.5	3.0	10.0
Pinhão	13	40.0	50.0	24.0	11.00	4.35	10.0	1.6	8.0
Trezói	8	50.0	62.5	30.0	13.75	8.00	12.5	3.0	10.0

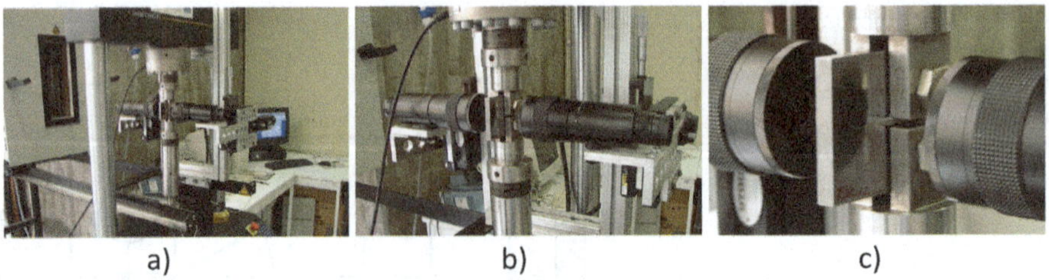

Fig. 5.52 Experimental fatigue crack growth tests: **a** overview of the *CT* specimen test; **b** two xy traveling microscopes measuring the crack advance on both faces; **c** two magnification microscopes on both faces of the *CT* specimen [17]

bridge (six from the diagonal and seven from the bracing) and eight specimens from the Trezói bridge.

The following stress ratios were investigated for each material:

- Eiffel bridge: $R_\sigma = 0.1$ and $R_\sigma = 0.5$;
- Luiz I bridge: $R_\sigma = 0.1$;
- Fão bridge: $R_\sigma = 0.0$, $R_\sigma = 0.25$, $R_\sigma = 0.5$ and $R_\sigma = 0.75$;
- Pinhão bridge: $R_\sigma = 0.0$, $R_\sigma = 0.1$ and $R_\sigma = 0.5$;
- Trezói bridge: $R_\sigma = 0.0$, $R_\sigma = 0.25$ and $R_\sigma = 0.5$.

The experimental crack propagation data obtained for each material related the crack propagation rate to the stress intensity factor range, using the power law as proposed by Paris and Erdogan [55]:

$$\frac{da}{dN} = C \cdot \Delta K^m \tag{5.8}$$

where *da/dN* is the fatigue crack propagation rate, ΔK represents the stress intensity factor range and *C* and *m* are material constants. The Paris's law can be modified to take into account the stress ratio effect using the Walker's modification [39]:

$$\frac{da}{dN} = C_1 \cdot \left[\frac{\Delta K}{(1 - R_\sigma)^{1-\gamma}}\right]^{m_1} \tag{5.9}$$

where C_1, m_1 and γ are constants to be determined from crack propagation data evaluated at distinct stress ratios. Using the equivalent stress intensity range defined as follows:

$$\overline{\Delta K} = \frac{\Delta K}{(1 - R_\sigma)^{1-\gamma}}, \tag{5.10}$$

a higher correlation will result for the fatigue crack propagation rates for distinct stress ratios. The formulation to determine the stress intensity factor range, ΔK, for the CT geometry is included in ASTM 647-99 standard [27], according to the following form:

$$\Delta K = \frac{\Delta F}{B\sqrt{W}} \cdot \frac{2+\alpha}{(1-\alpha)^{3/2}}(0.886 + 4.64\alpha - 13.32\alpha^2 + 14.72\alpha^3 - 5.6\alpha^4) \tag{5.11}$$

where: $\alpha = a/W$; a is the corresponding crack size, measured from the line of application of the load; ΔF is the force range; B and W are the thickness and nominal width of the CT specimens, respectively. The relation (5.11) is valid for $a/W \geq 0.2$.

For the MT geometry, the value of stress intensity factor range, ΔK, is obtained using the following expression:

$$\Delta K = \frac{\Delta F}{B} \cdot \sqrt{\frac{\pi \cdot \alpha}{2W} \sec \frac{\pi \cdot \alpha}{2}} \tag{5.12}$$

where: $\alpha = 2a/W$. The relation (5.12) is valid for $2a/W < 0.95$.

The main result of the fatigue crack propagation tests documented in this section, corresponds to the fatigue crack propagation rates as a function of the stress intensity factor ranges. The fatigue crack propagation rates were determined using the incremental polynomial method, as stated in ASTM E647-99 standard [27]. This method is based on the adjustment of 2nd degree polynomials to successive sets of experimental data points, which define successive lengths of the crack as a function of the stress intensity factor range. The fatigue crack propagation rate results from the derivative of these 2nd degree polynomials, which are expressed as a function of the stress intensity factor range.

5.5.2 *Experimental Fatigue Crack Propagation Results*

In this section, a series of graphs are presented for various stress ratios [11–13, 15, 17, 18, 28], showing the evolution of the fatigue crack propagation rates as a function of the stress intensity factor range, for the materials extracted from the Eiffel, Luiz I, Fão, Pinhão and Tezói bridges. All fatigue crack propagation data was correlated with the Paris's law. Consequently, each graph includes the Paris constants as well as the respective determination coefficient.

5.5.2.1 Material from Eiffel Bridge

Figures 5.53 and 5.54 present the fatigue crack propagation rates of the material from the Eiffel bridge, for the two stress ratios tested, namely, $R_\sigma = 0.1$ and $R_\sigma = 0.5$ [11, 18]. For this bridge, crack propagation rates were measured in the girder longitudinal (L) direction, whereas only one test was carried out in the transverse direction (T). The crack propagation rates in the transverse direction showed a very irregular pattern. Figure 5.55 presents all fatigue crack propagation rates together, for both stress ratios ($R_\sigma = 0.1 + R_\sigma = 0.5$) form the material from the Eiffel bridge.

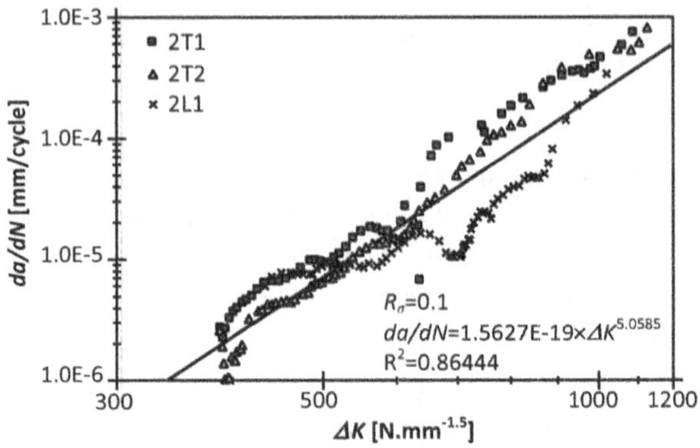

Fig. 5.53 Fatigue crack growth data of the material from the Eiffel bridge, $R_\sigma = 0.1$

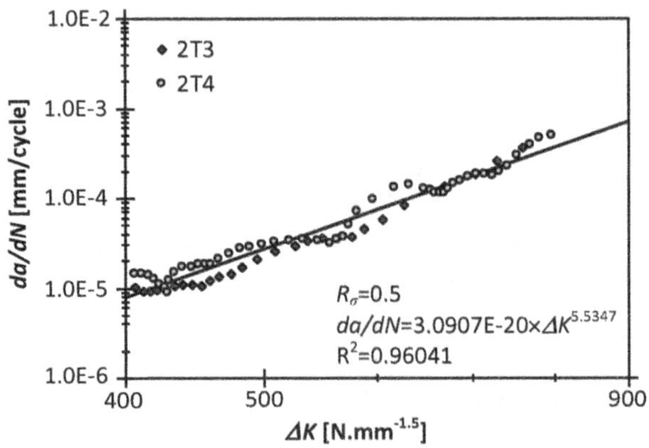

Fig. 5.54 Fatigue crack growth data of the material from the Eiffel bridge, $R_\sigma = 0.5$

Figure 5.56 compares the Paris mean regression lines for each stress ratio and that resulting from the aggregation of all stress ratios [11, 18]. The analysis of the figure reveals a clear stress ratio dependency of the fatigue crack propagation rates. A significant increase in the fatigue crack growth rates is observed for increasing stress ratios from 0.1 to 0.5.

5.5.2.2 Material from Luiz I Bridge

Figure 5.57 exhibits the fatigue crack propagation rates of the material from the Luiz I bridge, for the tested stress ratio, $R_\sigma = 0.1$ [11]. Data for only one stress ratio is available for this material. The scatter level is significant as pointed out by the relative low determination coefficient.

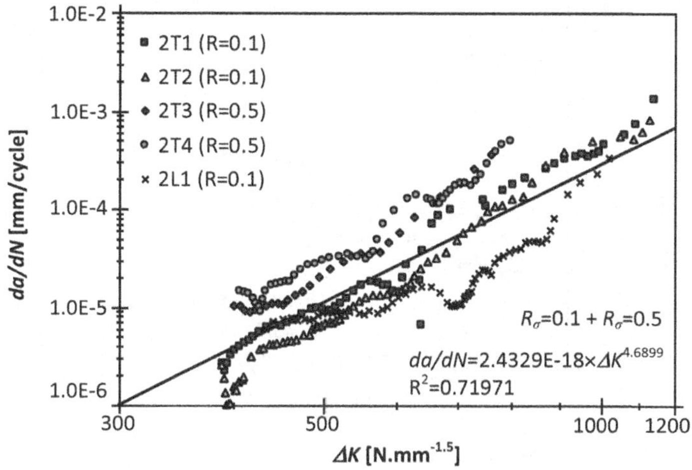

Fig. 5.55 Fatigue crack growth data of the material from the Eiffel bridge, $R_\sigma = 0.1 + R_\sigma = 0.5$

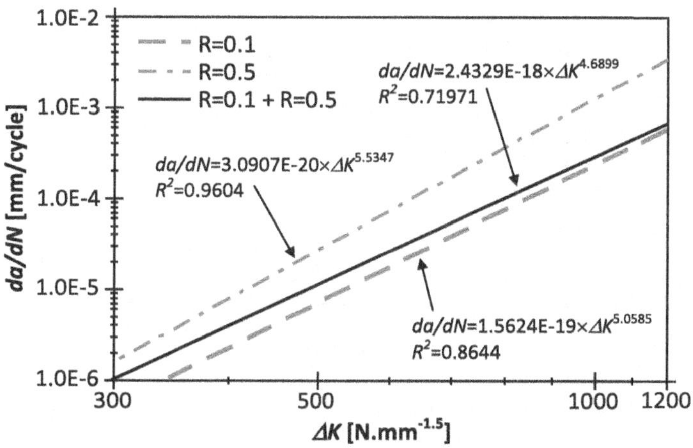

Fig. 5.56 Comparison between the regression lines of the fatigue crack propagation rates for different stress ratios obtained from the material of the Eiffel bridge

5.5.2.3 Material from Fão Bridge

Figures 5.58, 5.59, 5.60 and 5.61 show the fatigue crack propagation rates of the material from the Fão bridge, for the four stress ratios tested, namely, $R_\sigma = 0.0$, $R_\sigma = 0.25$, $R_\sigma = 0.5$ and $R_\sigma = 0.75$ [13]. Figure 5.62 presents together the fatigue crack propagation rates of the material from the Fão bridge for all stress ratios tested ($R_\sigma = 0.0 + R_\sigma = 0.25 + R_\sigma = 0.5 + R_\sigma = 0.75$). Comparing the determination coefficient from the global fitting of the Paris law with the determination coefficient of the individual fitting for each stress ratio, we realize a reduction in the determination

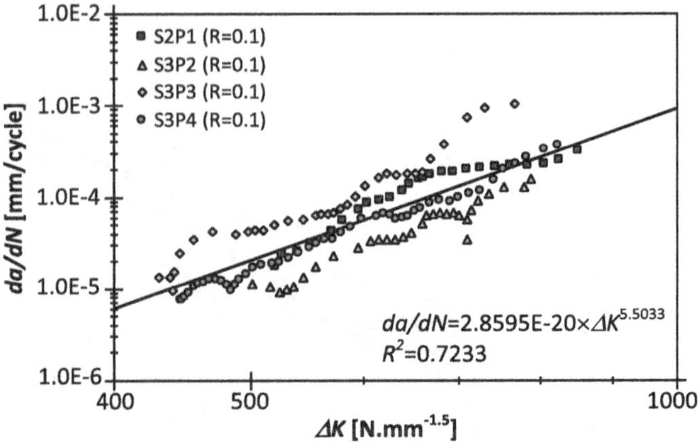

Fig. 5.57 Fatigue crack growth data for the material from the Luiz I bridge, $R_\sigma = 0.1$

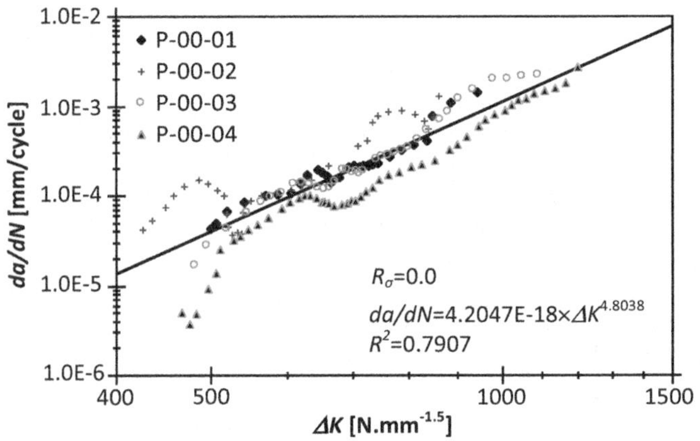

Fig. 5.58 Fatigue crack growth data for the material from the Fão bridge, $R_\sigma = 0.0$

coefficient, which is attributed to the stress ratio effects not being captured by the Paris relation. Figure 5.63 compares the average fatigue crack propagation rates, given by the Paris relation, for several stress ratios tested. The analysis of the graph reveals that the stress ratio influences the fatigue crack propagation rates. In general, an increase in the fatigue crack propagation rates is observed for increasing stress ratios [13, 18]. This phenomenon may be explained by the crack closure phenomenon occurring for lower stress ratios.

5.5.2.4 Material from Pinhão Bridge

Figures 5.64, 5.65 and 5.66 present the fatigue crack propagation rates for the material

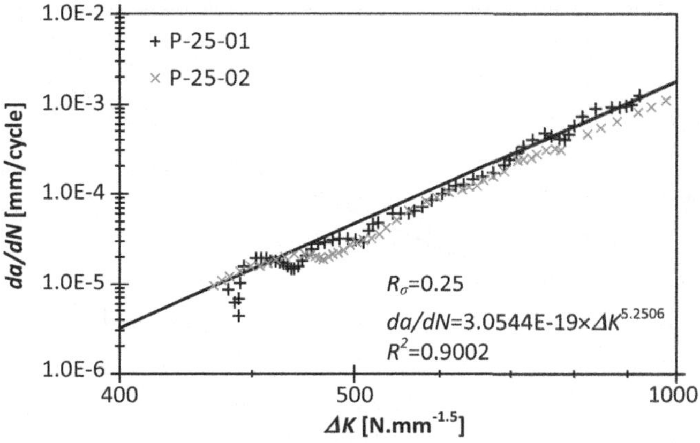

Fig. 5.59 Fatigue crack growth data for the material from the Fão bridge, $R_\sigma = 0.25$

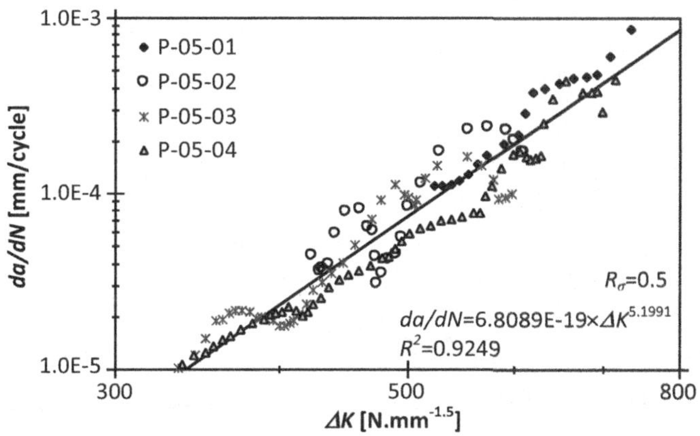

Fig. 5.60 Fatigue crack growth data for the material from the Fão bridge, $R_\sigma = 0.5$

from the Pinhão bridge, for three stress ratios, namely, $R_\sigma = 0.0$, $R_\sigma = 0.1$ and $R_\sigma = 0.5$ [15, 18]. Figure 5.67 shows the fatigue crack propagation data for all the stress ratios considered ($R_\sigma = 0.0 + R_\sigma = 0.1 + R_\sigma = 0.5$). Figure 5.68 compares the trend lines of the fatigue crack propagation rates for each stress ratio. The analysis of the figure proves an increase of the fatigue crack propagation rates for increasing stress ratio, mainly for lower stress intensity factor ranges while for higher stress intensity factor ranges, the fatigue crack propagation rates tend to be less sensitive to the stress ratio.

The two samples from the material of the Pinhão bridge, namely those from the bracing (B) and the diagonal (C), exhibit slightly distinct crack growth rates for $R_\sigma = 0.0$; however, for the other stress ratios no differences for the crack growth rates have been found between these two material samples.

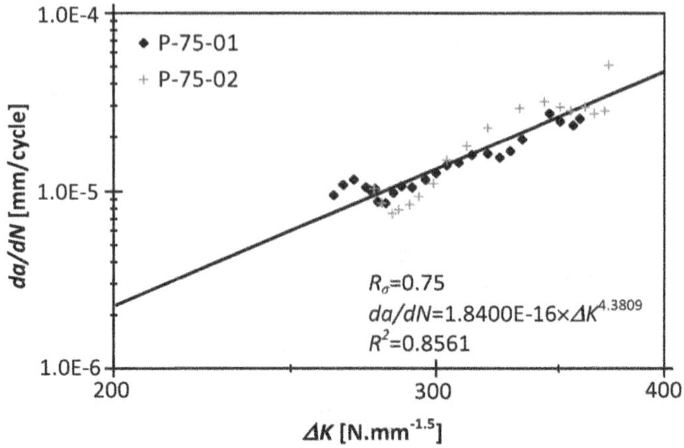

Fig. 5.61 Fatigue crack growth data for the material from the Fão bridge, $R_\sigma = 0.75$

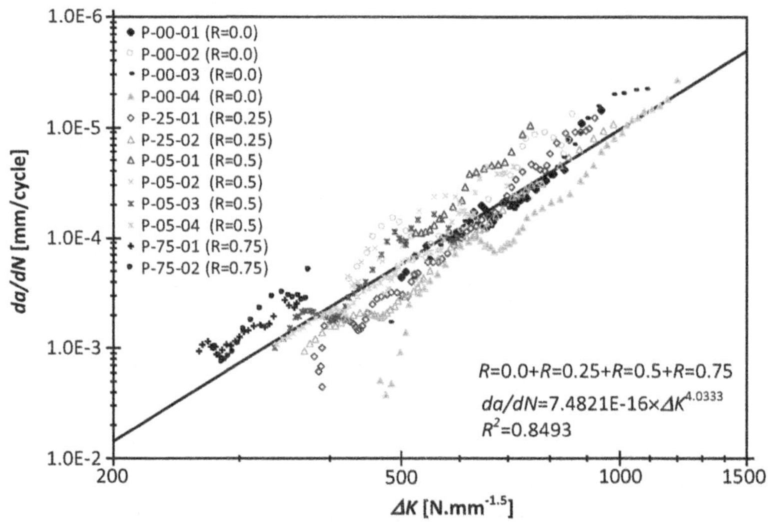

Fig. 5.62 Fatigue crack growth data for the material from the Fão bridge, $R_\sigma = 0 + R_\sigma = 0.25 + R_\sigma = 0.5 + R_\sigma = 0.75$

5.5.2.5 Material from Trezói Bridge

Figures 5.69, 5.70 and 5.71 present the fatigue crack propagation rates of the material from the Trezói bridge, for three tested stress ratios, namely $R_\sigma = 0.0$, $R_\sigma = 0.25$ and $R_\sigma = 0.5$ [17, 18]. Figure 5.72 illustrates the fatigue crack propagation rates for the material from the Trezói bridge for all stress ratios joined together. This material shows clearly a reduced scatter in the crack propagation rates. This material is a modern construction steel and therefore, shows less scatter in mechanical properties

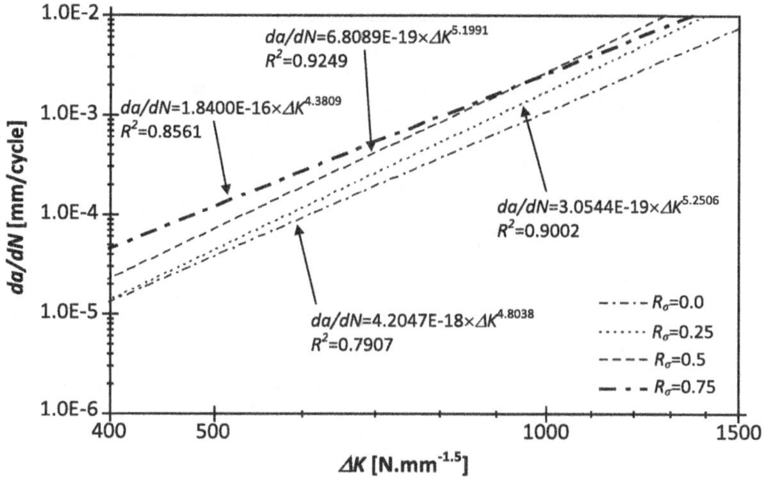

Fig. 5.63 Comparison among the regression lines of the fatigue crack propagation rates for different stress ratios obtained from the material of the Fão bridge

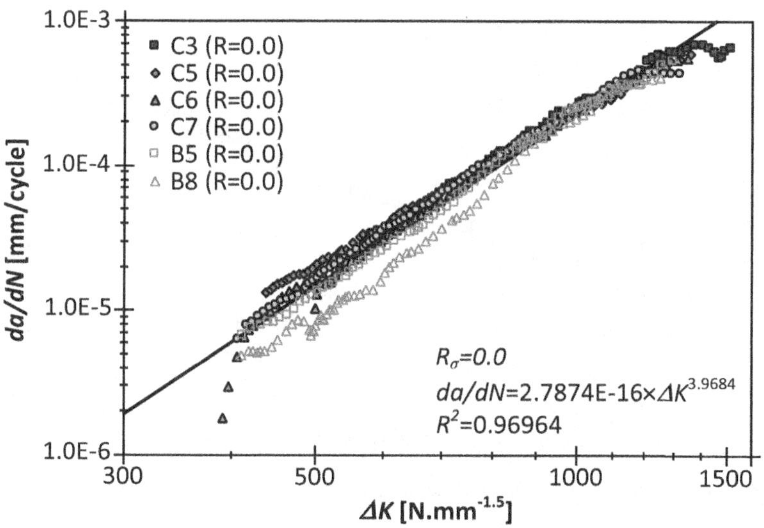

Fig. 5.64 Fatigue crack growth data for the material from the Pinhão bridge, $R_\sigma = 0.0$

than old puddle irons. Figure 5.73 compares the trends in the fatigue crack propagation rates of the material for the stress ratios tested. It is visible that this material also shows some dependency with the stress ratio, mainly when stress ratio changes between $R_\sigma = 0.0$ and $R_\sigma = 0.25$ on the contrary, for stress ratios changing between $R_\sigma = 0.25$ and $R_\sigma = 0.5$, no significant variation of the fatigue crack growth rates

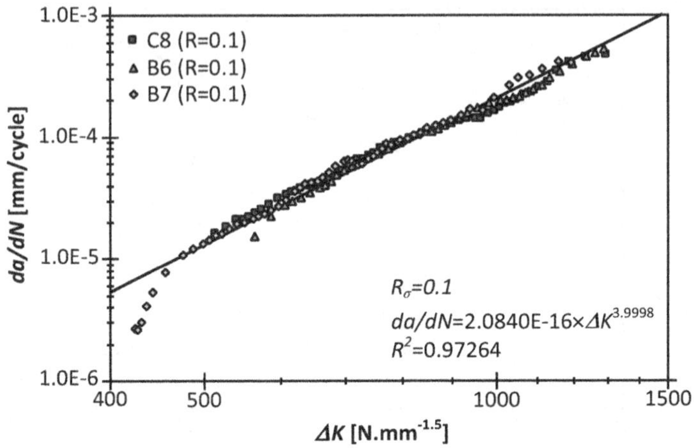

Fig. 5.65 Fatigue crack growth data for the material from the Pinhão bridge, $R_\sigma = 0.1$

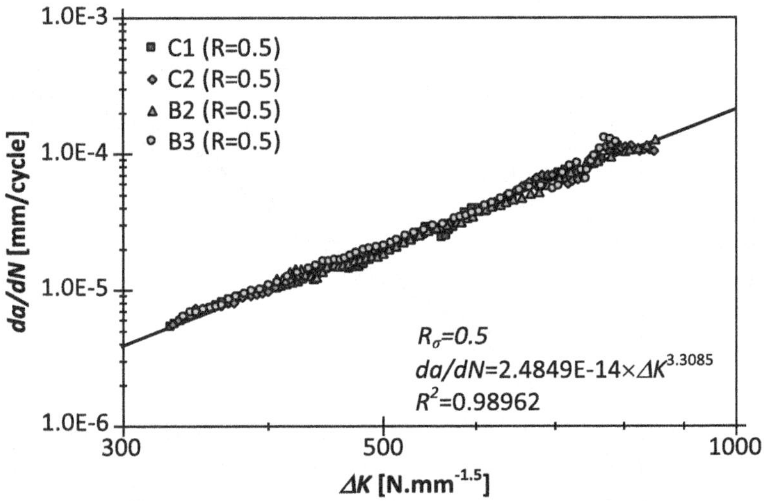

Fig. 5.66 Fatigue crack growth data for the material from the Pinhão bridge, $R_\sigma = 0.5$

are observed. This means that any possible crack closure that may occur will vanish above $R_\sigma = 0.25$.

5.5.2.6 Discussion of the Results

The greatest scatter in fatigue crack propagation data was found for the materials from Eiffel, Luiz I and Fão bridges which is consistent with the fact that these bridges

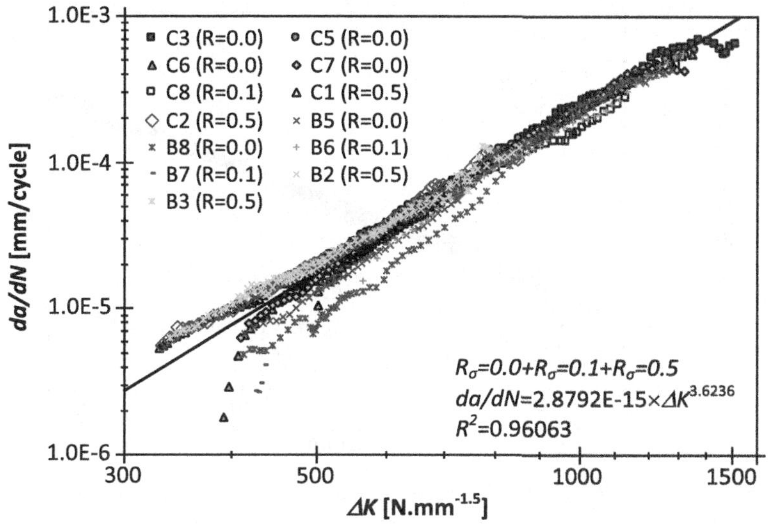

Fig. 5.67 Fatigue crack growth data for the material from the Pinhão bridge, $R_\sigma = 0 + R_\sigma = 0.1 + R_\sigma = 0.5$

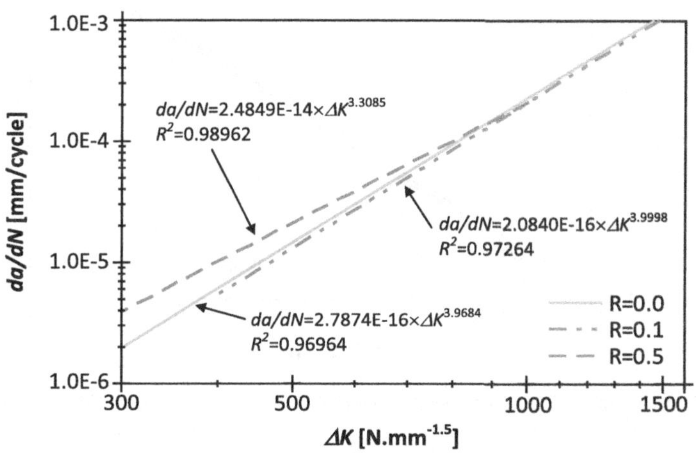

Fig. 5.68 Comparison between the regression lines of the fatigue crack propagation rates for different stress ratios obtained from the material of the Pinhão bridge

are the oldest ones and the respective materials present important heterogeneities. The data from Pinhão bridge reveals a relative low scatter; this material being about 25 years younger than the previous, one revealing good homogeneity, similar to modern steels. Finally, the material from the Trezoi bridge exhibits very low scatter, as expected, due to the relative low age of the material. However, there is one exception with the material from the Trezói bridge, which shows lower determination coefficient

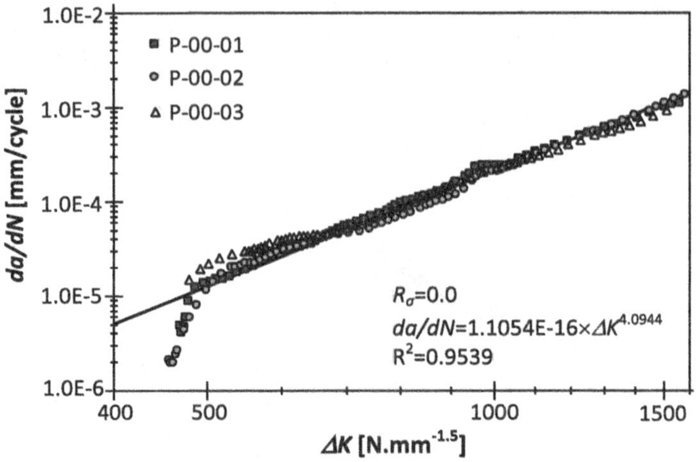

Fig. 5.69 Fatigue crack growth data for the material from the Trezói bridge, $R_\sigma = 0.0$

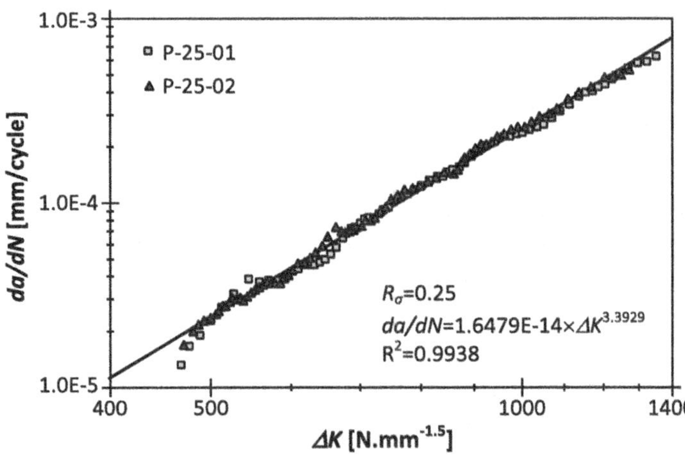

Fig. 5.70 Fatigue crack growth data for the material from the Trezói bridge, $R_\sigma = 0.25$

than that from the Pinhão bridge. This apparent exception may be justified by the fact that data from the Trezói bridge for $R_\sigma = 0$ covers the near threshold crack propagation, which is not reproduced by the Paris's law.

For the material from the Eiffel bridge (Figs. 5.53, 5.54, 5.55 and 5.56), it is clear that crack propagation increases for increasing stress ratio effect, which may be justified by the reduction of the crack closure effect. The crack propagation rate of the material from the Eiffel bridge was the most sensitive to the stress ratio. Regarding the material from the Luiz I bridge (Fig. 5.57), only one stress ratio was tested ($R_\sigma = 0.1$), so that it was not possible to assess the stress ratio effects on the fatigue crack propagation for this material. Four stress ratios were tested for the sample of material

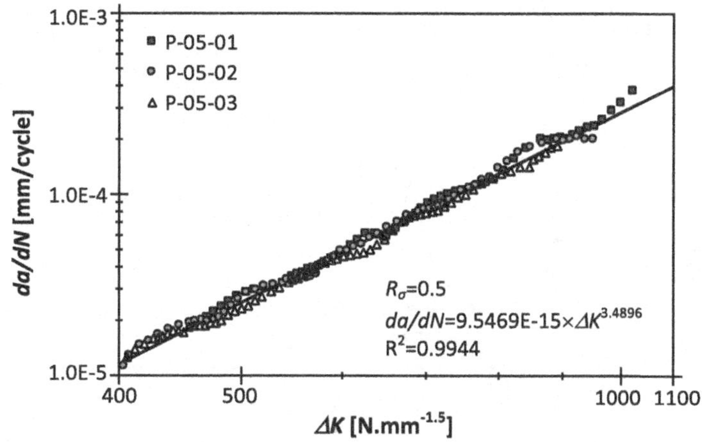

Fig. 5.71 Fatigue crack growth data for the material from the Trezói bridge, $R_\sigma = 0.5$

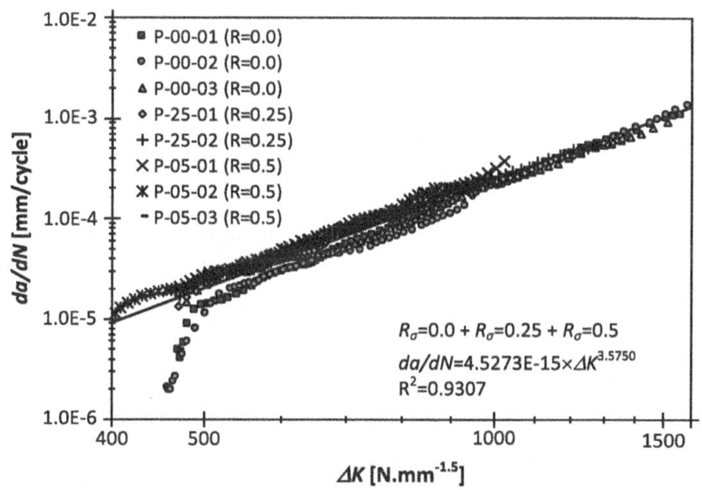

Fig. 5.72 Fatigue crack growth data for the material from the Trezói bridge, $R_\sigma = 0 + R_\sigma = 0.25 + R_\sigma = 0.5$

from the Fão bridge (Figs. 5.58, 5.59, 5.60, 5.61, 5.62 and 5.63), namely $R_\sigma = 0$, $R_\sigma = 0.25$, $R_\sigma = 0.5$ and $R_\sigma = 0.75$. For this material a slight increase of the fatigue crack propagation rate is also observed for increasing stress ratio. Finally, the materials from the Pinhão (Figs. 5.64, 5.65, 5.66, 5.67 and 5.68) and Trezói (Figs. 5.69, 5.70, 5.71, 5.72 and 5.73) bridges showed a slight acceleration in the crack propagation for increasing stress ratio.

It is interesting to note that the exponent of the Paris's law [29], m, is always higher than 3 (value adopted in design codes) and falls in the range 3.6–5.5. The coefficient C is within the range 2.9×10^{-15}–2.9×10^{-20} (da/dN in mm/cycle and

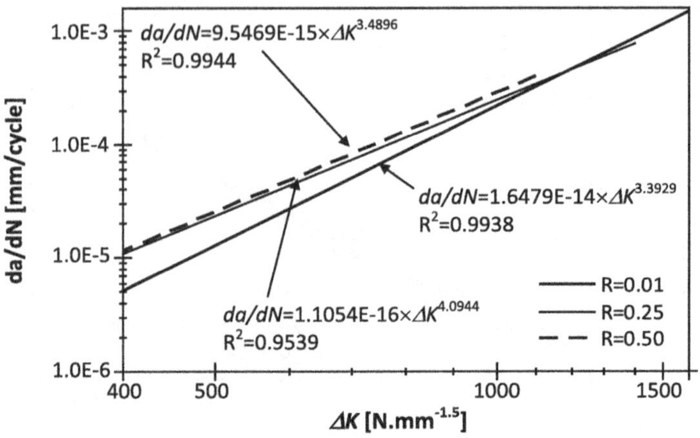

Fig. 5.73 Comparison between fatigue crack propagation rate trends for different stress ratios obtained for the material of the Trezói bridge

ΔK in $Nmm^{-1.5}$). Table 5.31 summarizes the material constants, *C* and *m*, of the Paris´s model for each bridge material.

Table 5.31 Constants of the Paris's model for each material of the bridges

Material	R_σ	No. specimens	C*	*m*	R^2
Eiffel	0.1	3	1.5627E-19	5.0585	0.8644
	0.50	2	3.0907E-20	5.5347	0.9604
	0.1; 0.5	5	2.4329E-18	4.6899	0.7197
Luiz I	0.1	4	2.8595E-20	5.5033	0.7233
Fão	0.0	4	4.2047E-18	4.8038	0.7907
	0.25	2	3.0544E-19	5.2506	0.9002
	0.5	4	6.8089E-19	5.1991	0.9249
	0.75	2	1.8400E-16	4.3809	0.8561
	0.0; 0.25; 0.5; 0.75	12	7.4821E-16	4.0333	0.8493
Pinhão	0.0	6	2.7874E-16	3.9684	0.9694
	0.1	3	2.0840E-16	3.9998	0.9726
	0.5	4	2.4849E-14	3.3085	0.9896
	0.0; 0.1; 0.5	13	2.8792E-15	3.6236	0.9606
Trezói	0.0	3	1.1054E-15	4.0944	0.9539
	0.25	2	1.6479E-14	3.3929	0.9938
	0.5	3	9.5469E-15	3.4896	0.9944
	0.0; 0.25; 0.5	8	4.5273E-15	3.5750	0.9307

*da/dN [mm/cycle] and DK [$N.mm^{-1.5}$]

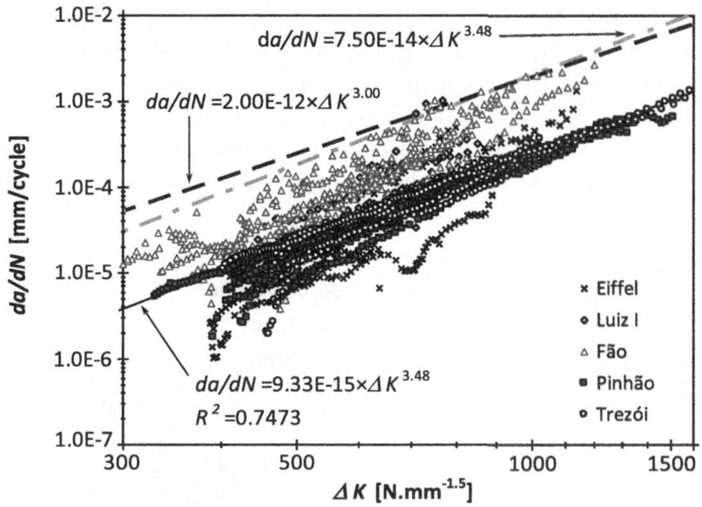

Fig. 5.74 Correlation among all fatigue crack propagation data using the Paris model

A tentative is done to derive a unique relation for the fatigue crack propagation data from all bridge materials, joined together. Figure 5.74 plots the fatigue crack propagation data obtained for all materials. The best fit of the Paris relation provides a determination coefficient, $R^2 = 0.75$, which is relatively high, taking into account the different origins of the materials investigated. It can be verified that fatigue data of the materials from Fão and Luiz I bridges diverge from the mean values, presenting higher fatigue crack propagation values. One specimen from Eiffel bridge exhibits markedly lower fatigue crack propagation rates for intermediate stress intensity factor ranges. The constant *m* from the Paris's law obtained for all data together, is higher than 3.0, this being the latter value usually adopted in current design codes of practice [5, 30], indirectly defined as the slope of the S–N curves of structural details. The constant *C* is significantly lower than the usual values recommended in the literature for modern steels [30]: $1.2 \times 10^{-13} \leq C \leq 5 \times 10^{-13}$. Figure Fig. 5.74 also includes an upper bound, parallel to the linear regression line, which can be used for design purposes. Another upper bound was established based on a slope of 3. For this latter case, the *C* constant happens to be higher than the values referred in literature for modern steels [30].

Figures 5.75, 5.76, 5.77 and 5.78 plot again the fatigue crack propagation data for the materials from the bridges investigated, but in these graphs the data is plotted by considering the effective stress intensity factor range, as proposed by Walker [18]. The material from the Luiz I bridge is excluded from this analysis since only data for $R_\sigma = 0.1$ is available. The application of the Walker model [39] to correlate the fatigue crack propagation data resulted in higher determination coefficients, when

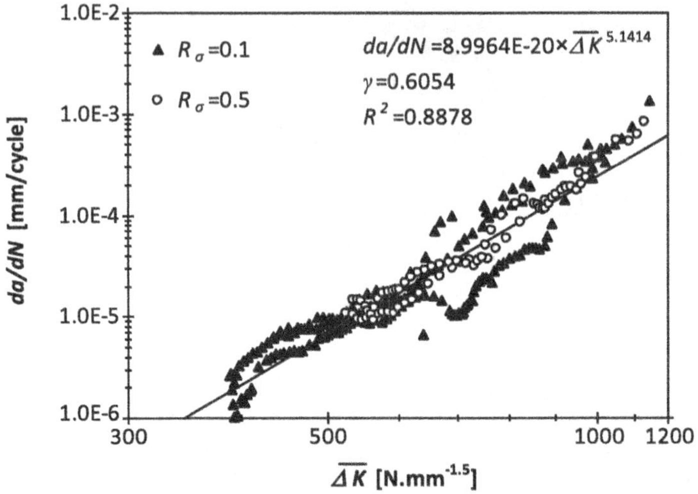

Fig. 5.75 Correlation between the fatigue crack propagation data from the material of the Eiffel bridge using Walker's model

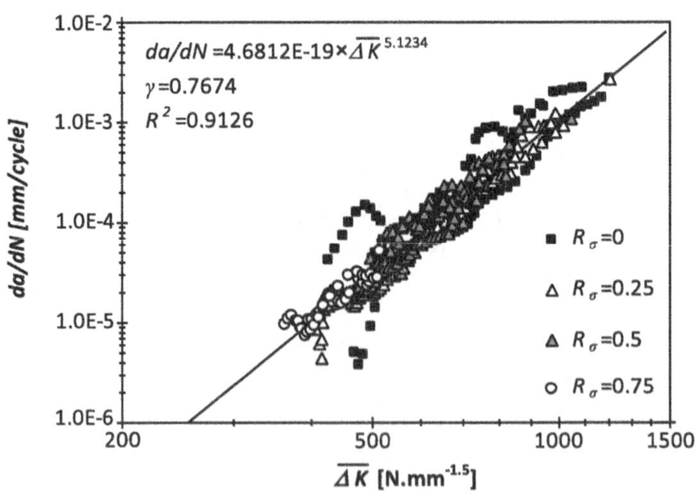

Fig. 5.76 Correlation between the fatigue crack propagation data from the material of the Fão bridge using Walker's model

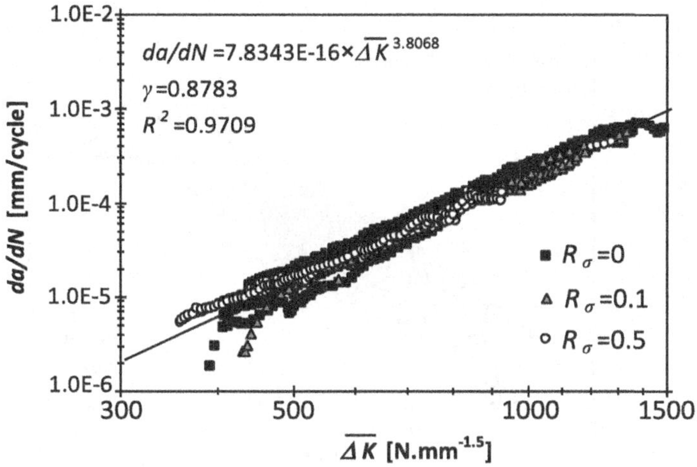

Fig. 5.77 Correlation for the fatigue crack propagation data from the material of the Pinhão bridge using Walker's model

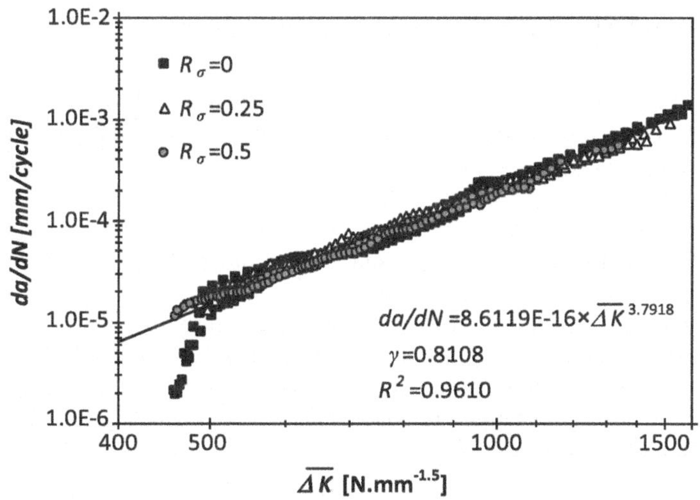

Fig. 5.78 Correlation between the fatigue crack propagation data from the material of the Trezói bridge using Walker's model

compared with those resulting when applying the Paris law. Figure 5.79 plots all data together, using the transformation proposed by Eq. 5.10. The data is correlated using a power relation, and an increase of the determination coefficient is again verified [18].

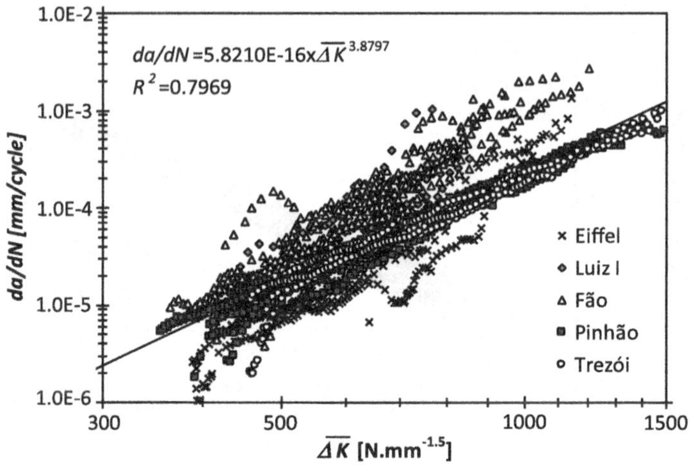

Fig. 5.79 Correlation between the fatigue crack propagation rates for all bridge materials using Walker's model

5.6 Fatigue Strength of Riveted Connections

The fatigue resistance of riveted joints made of original materials from the Portuguese bridges has been evaluated by means of fatigue tests [12, 14–17]. Riveted joints with original rivets were extracted from the Eiffel, Luiz I, Pinhão and Trezói bridges. In addition, riveted joints were manufactured using original material from the Fão bridge, but applying new rivets. In this section, a summary of available S-N fatigue data, from riveted connections representative of old metallic riveted Portuguese bridges, is presented.

The results of the fatigue tests of riveted joints are presented in the form of S-N or Wöhler curves, which shows a power relationship between the applied stress range ($\Delta\sigma$) and the corresponding number of cycles to failure (N_f) for the finite life domain and excluding low-cycle fatigue regimes. Design codes of practice [31] adopt the following form for the S-N curves:

$$\Delta\sigma^m N_f = C \tag{5.13}$$

where C and m are constants. Alternatively, the mean S-N curves may be represented using the following linear model [32, 33], which is more suitable for parameter identification using linear regression analysis:

$$Y = A + B \cdot X \tag{5.14}$$

where Y is the dependent variable defined as $Log(N_f)$, X is the independent variable defined as $Log(\Delta\sigma)$, A and B are linear regression parameters. Consequently, it is possible to rewrite the $S\text{-}N$ curve in the following forms [32, 33]:

$$\begin{cases} Log(N_f) = A + B \cdot Log(\Delta\sigma) \\ Log(\Delta\sigma) = -\frac{A}{B} + \frac{1}{B} Log(N_f) \end{cases} \tag{5.15}$$

where A and B parameters are related to the C and m constants:

$$\begin{cases} C = 10^A \\ m = -B \end{cases} \tag{5.16}$$

The S-N results presented in this section, were obtained using fatigue tests of specimens, performed on servo-hydraulic machines rated to 100 kN or 250 kN and subjected to load control conditions. The results of the fatigue tests of riveted connections for the materials under consideration from the Eiffel, Luiz I, Fão, Pinhão and Trezói bridges are summarized, respectively in Table 5.32, Table 5.33, Table 5.34, Table 5.35 and Table 5.36. In those tables S_{gross} is the gross cross-section, S_{net} is the net cross-section, R_σ is the stress ratio, F_{max} is the maximum testing force, ΔF is the test load range, f is the test frequency, $\Delta\sigma_{gross}$ is the stress range computed on the gross cross-section, $\Delta\sigma_{net}$ is the stress range calculated in the net cross-section and N_f is the number of cycles to failure [12, 14–17, 28]. The gross cross-section corresponds to the remote section of the joint, while the net cross-section corresponds to the section excluding the rivet hole.

5.6.1 *Riveted Connections from the Eiffel Bridge*

Double shear riveted connections were cut out from the web of a girder from the Darque south viaduct of the Eiffel bridge. The specimens were cut along the longitudinal direction and the respective mid plates showed an average thickness of 6.7 mm. Figure 5.80 illustrates the complete test series of the riveted specimens extracted from the Eiffel bridge. A total of 14 specimens were tested under a stress R-ratio equal to 0.1 and frequencies, f, ranged between 5 and 7.5 Hz. Each specimen exhibits a single row of four rivets, aligned in the loading direction. The results of these fatigue tests are summarized in Table 5.32.

5.6.2 *Riveted Connections from the Luiz I Bridge*

Original single shear riveted specimens with single rivets were tested from the Luiz I bridge, the geometry of which is illustrated in Fig. 5.80 [12]. Due to limitations imposed by the available material, only single lap joints were tested. A total of 7 specimens were tested under $R_\sigma = 0.1$ and test frequencies, f, ranged between 6 and 8 Hz. The respective results are summarized in Table 5.33 [12] (Fig. 5.81).

Table 5.32 Results of the fatigue tests of the riveted joints from the Eiffel bridge

Specimen	S_{gross} mm^2	S_{net} mm^2	R_σ	F_{max} N	ΔF N	f Hz	$\Delta\sigma_{gross}$ MPa	$\Delta\sigma_{net}$ MPa	N_f cycles
V1	430.1	296.8	0.1	55533	49980	5.0	116.2	168.4	1513265
V2	427.3	293.5		54939	49445	7.5	115.7	168.4	1500331
V3	445.3	278.7		57538	51784	7.5	116.3	185.8	240383
V4	432.0	299.6		55640	50076	7.5	115.9	167.1	149378
V5	433.7	303.3		67551	60796	5.0	140.2	200.5	61456
V6	442.2	301.2		69239	62315	5.0	140.9	206.9	149879
V7	427.9	280.4		66115	59504	5.0	139.1	212.2	19768
V8	437.8	308.1		68511	61660	5.0	140.8	200.2	12195
V9	431.2	304.3		61121	55009	5.0	127.6	180.8	31954
V10	438.1	306.9		62122	55910	5.0	127.6	182.1	53638
V11	441.4	285.9		63062	56756	5.0	128.6	198.5	114510
V12	452.8	311.2		64352	57917	5.0	127.9	186.1	23196
V13	428.5	259.0		54480	49032	5.0	114.4	189.3	91069
V14	431.1	309.3		66909	60218	5.0	139.7	194.7	12950

Table 5.33 Results of the fatigue tests of the riveted joints from the Luiz I bridge [12]

Specimen	S_{gross} mm^2	S_{net} mm^2	R_σ	F_{max} N	ΔF N	f Hz	$\Delta\sigma_{gross}$ MPa	$\Delta\sigma_{net}$ MPa	N_f cycles
S5R1	350.0	210.0	0.1	31500	28350	8.0	81.0	135.0	164985
S5R2	350.0	210.0		31500	28350	6.0	81.0	135.0	426259
S5R3	350.0	210.0		26600	23940	7.0	68.4	114.0	161801
S5R4	350.0	210.0		22600	20340	7.0	58.1	96.9	999453
S5R5	350.0	210.0		19200	17280	7.0	49.4	82.3	514569
S5R6	350.0	210.0		16330	14697	8.0	42.0	70.0	1586560
S5R7	350.0	210.0		14000	12600	8.0	36.0	60.0	904586

Table 5.34 Results of the fatigue tests for riveted joints from the Fão bridge [14]

Specimen	S_{gross} mm^2	S_{net} mm^2	R_σ	F_{max} N	ΔF N	f Hz	$\Delta\sigma_{gross}$ MPa	$\Delta\sigma_{net}$ MPa	N_f cycles
FA1	342.2	160.2	0.01	56940	56370.6	2.5	164.7	351.9	9744
FA2	345.7	163.8		58210	57627.9	2.5	166.7	351.9	5285
FA3	342.9	162.3		57690	57113.1	2.5	166.5	351.9	24357
FA4	340.9	160.3		56960	56390.4	5.0	165.4	351.9	3458
FA5	333.6	157.8		49300	48807	2.5	146.3	309.3	141330
FA6	329.4	155.5		48570	48084.3	5.0	146.0	309.3	52879
FA7	349.2	165.6		51730	51212.7	5.0	146.6	309.3	6644
FA8	345.6	163.8		51160	50648.4	5.0	146.5	309.2	21050
FA9	344.3	163.1		47430	46955.7	5.0	136.4	287.9	38242
FA10	344.0	163.4		47530	47054.7	5.0	136.8	287.9	103809
FA11	345.9	163.1		47430	46955.7	5.0	135.8	287.9	75749
FA12	343.7	163.0		35120	34768.8	5.0	101.2	213.3	210995
FA13	327.7	155.3		26770	26502.3	5.0	80.9	170.7	699161
FA14	347.2	160.8		27700	27423	10.0	79.0	170.6	4000000→
FA15	338.7	159.7		27520	27244.8	5.0	80.5	170.6	1439414

Table 5.35 Results of the fatigue tests of riveted joints from the Pinhão bridge [15]

Specimen	S_{gross} mm^2	S_{net} mm^2	R_σ	F_{max} N	ΔF N	f Hz	$\Delta\sigma_{gross}$ MPa	$\Delta\sigma_{net}$ MPa	N_f cycles
CF1	460.3	269.3	0.1	25000	22500	6.0	48.9	83.5	1922024
CF2	458.4	267.4		50000	45000	6.0	98.2	168.3	86140
CF3	459.4	268.4		37000	33300	6.0	72.5	124.1	635172
CF4	460.3	269.3		31000	27900	6.0	60.6	103.6	574452
CF5	460.3	269.3		18500	16650	6.0	36.2	61.8	1450789
CF6	460.3	269.3		25000	22500	6.0	48.9	83.5	2243676
CF7	460.3	269.3		18500	16650	6.0	36.2	61.8	10000000→

Table 5.36 Results of the fatigue tests for riveted joints from the Trezói bridge [16, 17]

Specimen	S_{gross} mm^2	S_{net} mm^2	R_σ	F_{max} N	ΔF N	f Hz	$\Delta\sigma_{gross}$ MPa	$\Delta\sigma_{net}$ MPa	N_f cycles
F1	591.3	328.5	0.1	71000	63900	4.0	108.1	194.5	50771
F2	580.5	322.5		47500	42750	6.0	73.6	132.6	605387
F3	585.9	325.5		47500	42750	6.0	73.0	131.3	566477
F4	495.8	235.4		30000	27000	10.0	54.5	114.7	2518224
F5	540.0	280.8		33000	29700	10.0	55.0	105.8	1202674
F6	498.1	236.5		36200	32580	8.0	65.4	137.8	846982
F7	499.1	239.7		29000	26100	10.0	52.3	108.9	4901965
F8	540.0	280.8		33000	29700	10.0	55.0	105.8	3473620

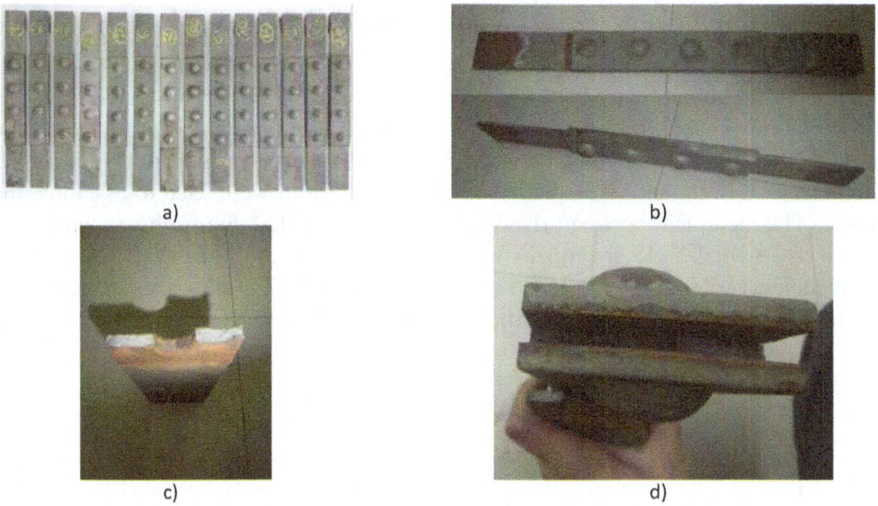

Fig. 5.80 Riveted specimens extracted from the Eiffel bridge: **a** complete test series; **b**, **c** and **d** illustration of the specimen, before and after fatigue failure

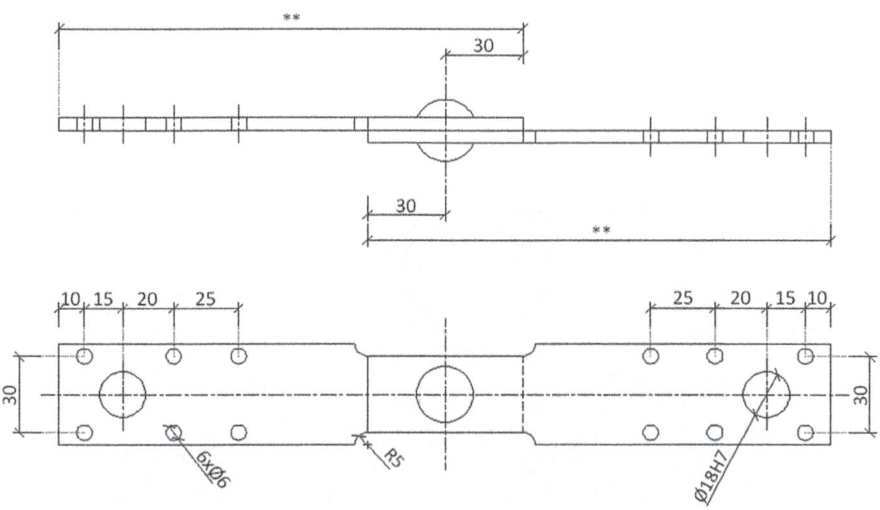

Fig. 5.81 Nominal geometry of the riveted joint from the Luiz I bridge [12]. **Maximum dimension allowing equal dimension of side plates

5.6.3 Riveted Connections from the Fão Bridge

Specimens manufactured with the original material from the Fão bridge were also of double shear type, as shown in Fig. 5.82. In this case, only two rivets were applied and the plate thickness ranged between 5 and 8 mm. New holes were drilled with a diameter of 24 mm and rivets with a diameter of 22 mm were used to assemble the plates. The clearance between rivets and holes were filled due to the expansion of the rivets. A total of 15 specimens were tested under stress ratio $R_\sigma = 0.0$ and test frequencies, f, ranging between 2.5 and 12 Hz. Table 5.34 summarizes the results of the fatigue tests [14] obtained for these riveted specimens.

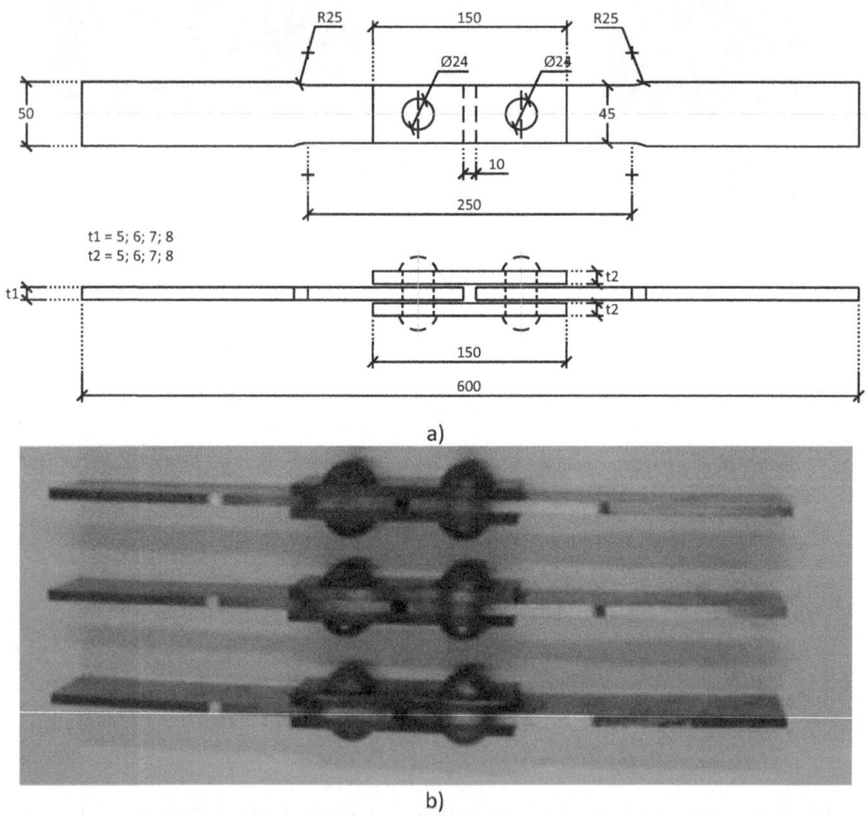

Fig. 5.82 Riveted joint prepared with the material from the Fão bridge: a) geometry of the riveted joint (dimensions in mm); b) views of the riveted specimens [14]

5.6.4 Riveted Connections from the Pinhão Bridge

Figure 5.83 illustrates the geometry of the original riveted specimens extracted from the Pinhão bridge, which was similar to that of the specimens extracted from the Luiz I bridge. An approximate thickness of the plates between 10 and 11 mm was measured for those specimens. The observation of a macrograph of the rivet longitudinal section allow us the estimation of the hole diameter (Ø21 mm) and the rivet diameter (Ø20 mm). A total of seven specimens were tested under load control, with stress R-ratio, $R_\sigma = 0.1$. The number of specimens was limited by the amount of available material. The results of the fatigue tests are summarized in Table 5.35 [15]. The fracture surfaces of specimens CF1, CF4 and CF5 show that the fatigue cracks initiated at existing flaws as illustrated in Fig. 5.84 [15].

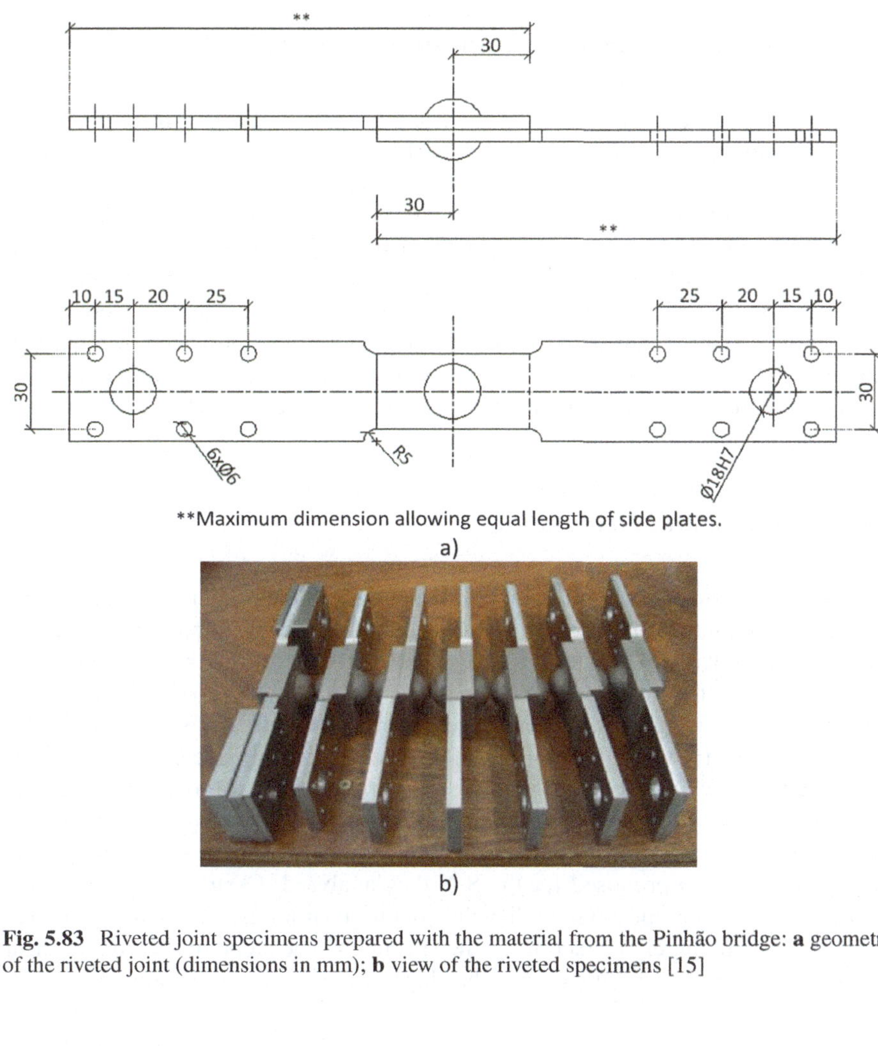

Fig. 5.83 Riveted joint specimens prepared with the material from the Pinhão bridge: **a** geometry of the riveted joint (dimensions in mm); **b** view of the riveted specimens [15]

Fig. 5.84 Initial crack-like flaws observed in some tested riveted specimens from the Pinhão bridge [15]

5.6.5 Riveted Connections from the Trezói Bridge

Riveted joints, with a single rivet and one shear plane, were machined from an original bracing member removed from the Trezói bridge. The final dimensions of the riveted joints are illustrated in Fig. 5.85 [16]. The original rivets of the connections were preserved. The riveted joints were fatigue tested under a stress R-ratio equal to 0.1 and test frequencies, f, ranging between 4 and 10 Hz. The S–N data is summarized in Table 5.36 [16, 17]. Some failures were generated at cracks that initiated at the rivet hole and propagated through the net cross section; other failures were due to cracks initiated and propagated outside the net cross section, motivated by important clamping forces on the rivets. The true value of the clamping forces in the rivets is, in general, difficult to assess and in this particular case were not measured.

5.6.6 Results and Discussion

A linear regression analysis was performed on the S-N experimental data available for the riveted joints from the old Portuguese metallic bridges. Figures 5.86 and 5.87 show the experimental S-N data and the respective S-N curves. The riveted joints were categorized into single and double shear joints. Relatively low coefficients resulted from the regression analysis, the lowest determination coefficient being obtained for the riveted joints from the Eiffel bridge. The parameters of the mean S-N curves are presented in Table 5.37, together with the stress range, $\Delta\sigma_c$, for a number of cycles to failure of 2×10^6 cycles. Besides the parameters of the individual S-N curves, corresponding to each tested riveted joint, the parameters for the single and double shear joints are also proposed for the S-N data analysed. Clearly, the single shear joints show lower fatigue strength than the double shear joints, which may be in part attributed to bending effects acting on the specimen. The slope of the double shear joints is also significantly higher than the slope of the single shear joints.

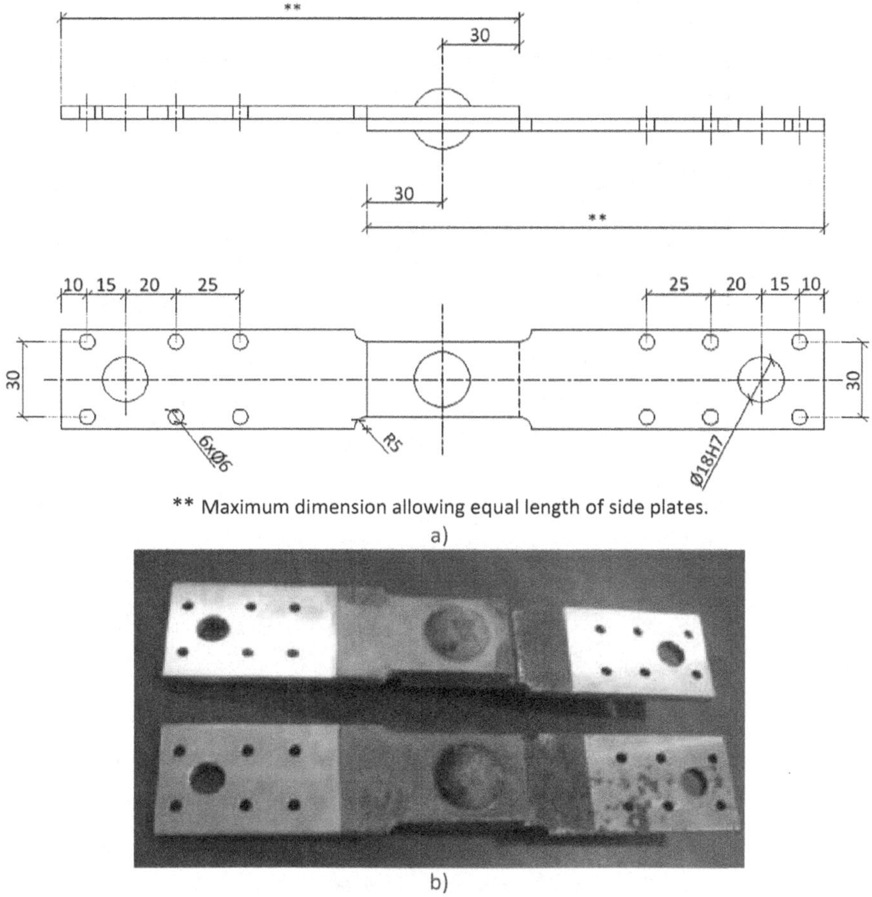

Fig. 5.85 Riveted joint prepared with material from the Trezói bridge: **a** geometry of the riveted joint (dimensions in mm); **b** view of the riveted specimens [16]

Taras and Greiner [34] have performed a statistical analysis of a significant amount of experimental fatigue data available in literature for riveted joints from old bridges. These authors suggested the categorization of the riveted joints into five categories. Two of the categories are the single and double shear splices under tensile loading. They also suggested the use of a slope, $m = 5$ for the design curve, instead of the $m = 3$ proposed in the design codes for the joints, EC3-1-9 [5]. In addition, they refer that mean stress effects must be accounted for riveted joints and suggested the use of a normalized stress range to allow the comparison of experimental fatigue data for distinct mean stresses to be made [34]:

$$\Delta\sigma_{nom} = \frac{\Delta\sigma}{f(R_\sigma)} \tag{5.17}$$

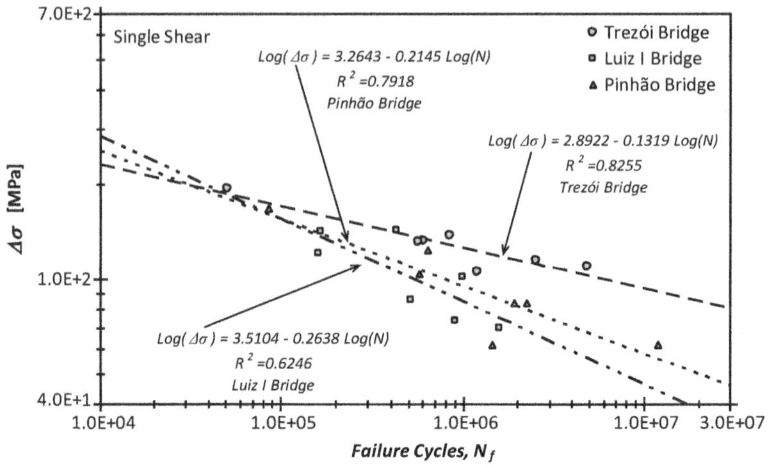

Fig. 5.86 *S-N* fatigue data for single shear riveted connections from Portuguese metallic bridges

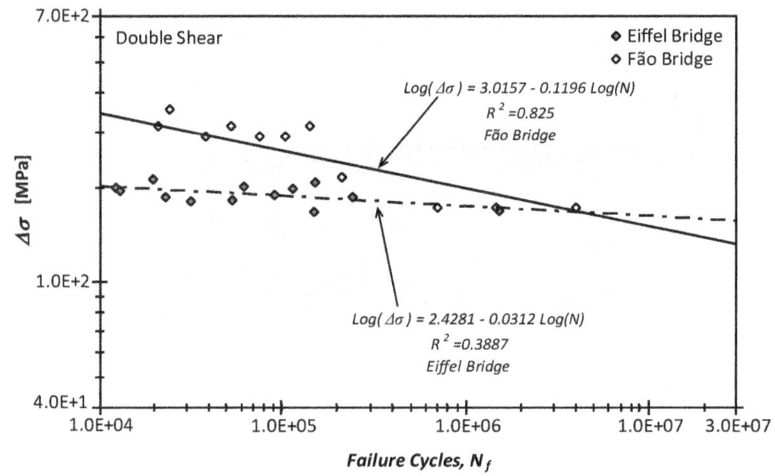

Fig. 5.87 *S-N* fatigue data for double shear riveted connections from Portuguese metallic bridges

Table 5.37 S-N curve parameters resulting from the experimental data

Bridge	Riveted joints	*C* MPa	m	$\Delta\sigma_C$
Trezói	Single shear	8.4570E + 21	7.5815	115.1
Pinhão		1.6527E + 15	4.6620	81.8
Luiz I		2.0279E + 13	3.7908	70.5
Fão	Double shear	1.6401E + 25	8.3612	182.9
Eiffel		6.6637E + 77	32.0513	170.4
Trezói, Pinhão and Luiz I	Single shear	1.2770E + 17	5.5310	89.9
Fão and Eiffel	Double shear	9.4512E + 28	10.2145	165.9

where $\Delta\sigma_{norm}$ is the normalized stress range, $\Delta\sigma$ is the tested stress range, $f(R_\sigma)$ is a normalization function to account for stress ratio effects, defined as a function of the material. For wrought iron and mild steel manufactured before 1900, $f(R_\sigma)$ is defined as [34]:

$$\begin{array}{c} f(R_\sigma) = \frac{1-R_\sigma}{1-0.70R_\sigma} \; -1 \le R_\sigma \le 0 \\ f(R_\sigma) = \frac{1-R_\sigma}{1-0.75R_\sigma} R_\sigma > 0 \end{array} \quad (5.18)$$

For mild steel after 1900 (St37, St48, St52, etc.) the following normalization function is proposed [34]:

$$\begin{array}{c} f(R_\sigma) = \frac{1-R_\sigma}{1-0.40R_\sigma} \; -1 \le R_\sigma \le 0 \\ f(R_\sigma) = \frac{1-R_\sigma}{1-0.60R_\sigma} R_\sigma > 0 \end{array} \quad (5.19)$$

Figures 5.88 and 5.89 compare the S-N data from riveted joints of Portuguese bridges with S-N data from Taras and Greiner investigation [34], with details categorization, respectively, into single and double joints. Also, the Eurocode 3, class 71 S-N curve is presented [5]. The S-N data is represented using the normalized stress function, as defined in Eq. 5.17. For double shear riveted joints, Taras and Greiner [34] proposed a design S-N curve with a slope, $m = 5$ and a fatigue strength of 90 MPa at 2×10^6 cycles, while for single shear riveted joints, a design S-N curve with a slope, $m = 5$ and a fatigue strength of 71 MPa at 2×10^6 cycles was proposed by these authors. Only very few points from Portuguese riveted bridges fall below these design S-N curves, namely corresponding to riveted connections from Eiffel (double shear), Luiz I and Pinhão bridges (single shear). This result may be justified

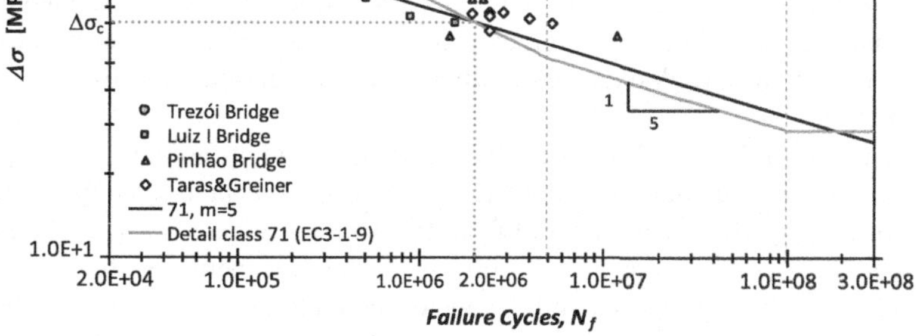

Fig. 5.88 Comparison of *S-N* data between Portuguese and international bridges details and proposal of a design *S-N* curve for single shear joints

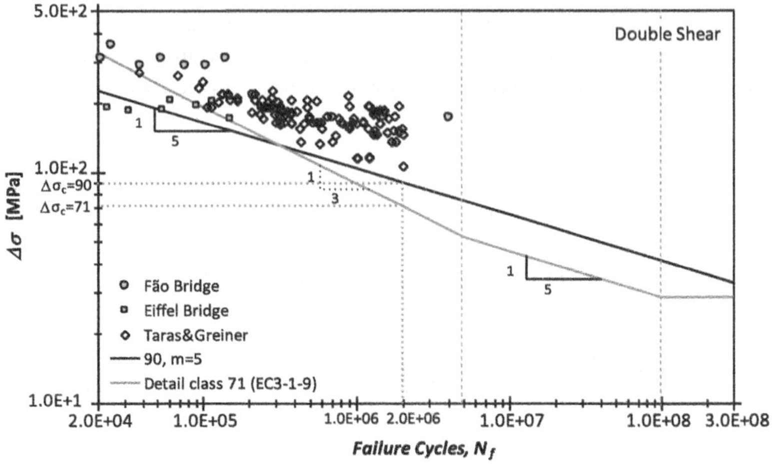

Fig. 5.89 Comparison of *S-N* data between Portuguese and international bridges details and proposal of a design *S-N* curve for double shear joints

by the high damage levels experienced by the riveted joints, due to the long bridge operation time. A higher slope, *m*, of the S-N curve is suggested by the experimental data, when a comparison is made with the class 71 S-N curve of the Eurocode EC3-1-9 [5]. Therefore, the proposition of new S-N curves based on a slope $m = 5$ seems to be an appropriate decision for riveted joints.

5.7 Concluding Remarks

In this chapter, strain-life and fatigue crack propagation data for samples of original materials removed from five Portuguese ancient riveted metallic bridges, namely from the Eiffel, Luiz I, Fão, Pinhão and Trezói bridges was compiled. Also, S-N fatigue resistance data from riveted joints made of original materials from the Portuguese bridges was gathered. The strain-life fatigue data was correlated using deterministic models. The fatigue crack propagation data was correlated using both the Paris model and the Walker model, the latter allowing the description of stress ratio effects.

Materials from the Eiffel, Luiz I, Fão and Pinhão bridges are very likely puddle irons account give of their age, the high microstructural heterogeneities and the low ductility properties. The material from the Trezói bridge is a low carbon structural steel. In general, there is a significant correlation between the mechanical performance, including fatigue, and the age of the materials.

Regarding the cyclic elastoplastic behaviour, the material from Trezói bridge does not follow the Masing behaviour while the materials from Eiffel and Luiz I bridge may be satisfactorily described by the Masing model, since some apparent deviation from the Masing behaviour may be justified by scatter in material properties. The material from Fão bridge shows a cyclic elastoplastic behaviour dependent on strain ratio, while the material from the Trezói bridge shows cyclic softening for strain ranges below 1%, the older materials show cyclic hardening.

Regarding the strain-life behaviour of the materials investigated, the number of transition reversals decreases with increasing age of the materials. The older materials show a transition life between low- and high-cycle fatigue regimes that is considered very low. Only the material from the Trezói bridge shows a transition life in the usual range for low carbon steels. Therefore, fatigue strength properties should play an important role on fatigue assessment of these old metallic bridges.

The fatigue crack propagation tests showed that the Paris law gives a good description of the fatigue crack growth data, for each stress ratio. The exponent of the Paris law resulted always greater than the value suggested by codes of practice ($m = 3$). The C coefficient was in order of magnitude lower than that recommended in literature for modern construction steels. A fatigue crack propagation design curve was proposed taking into account 42 fatigue crack propagation tests on original samples of the material from five distinct bridges. The Walker model improved the correlation of the fatigue crack propagation rates with the stress ratio effects.

S-N fatigue data from original riveted joints was compared with existing design curves. This comparison shows some data falling below the design S-N curves that corroborates the presence of cracks originated by the prior operation of the bridge or material degradation. The comparison performed with literature S-N data suggests the need of a riveted joint categorization, in particular the consideration of single and symmetric double shear splices. Also, a slope $m = 5$ seems to be more appropriate for riveted joints, rather than $m = 3$, as suggested by the current design codes. The influence of the stress R-ratio on S-N data for riveted joints has also to be accounted for in a convenient way.

Fracture mechanics and S-N based fatigue approaches have been used to assess riveted details, consisting on more classical approaches. Recent research suggests the use of local approaches to fatigue, supported on detailed finite element models [35–37]. The availability of adequate material properties is crucial. This issue is a limitation of the local approaches when materials under consideration are not fully investigated, and their fatigue properties are, consequently, not available.

References

1. Akesson B (1994) Fatigue life of riveted railway bridges. PhD thesis, Chalmers University of Technology, Sweden
2. Crocetti R (2001) On some fatigue problems related to steel bridges. PhD thesis, Chalmers University of Technology, Sweden
3. Mohammad AE (2002) Fatigue in riveted railway bridges. PhD thesis, Chalmers University of Technology, Sweden
4. Kühn B, Lukic´ M, Nussbaumer A, Günther H-P, Helmerich R, Herion S, et al (2008) In: Sedlacek G, Bijlaard F, Géradin M, Pinto A, Dimova S (eds) Assessment of existing steel

structures: recommendations for estimation of remaining fatigue life—EUR 23252 EN. JRC—ECCS
5. CEN-TC 250 (2003) EN 1993-1-9: Eurocode 3, Design of steel structures—Part 1-9: Fatigue. European Committee for Standardization, Brussels
6. AASHTO (1995) AASHTO LRFD: bridge design specification
7. De Jesus AMP, Figueiredo MAV, Ribeiro AS, de Castro PMST, Fernandes AA (2011) Residual lifetime assessment of an ancient riveted steel road bridge. Strain 47(S1):e402–e415
8. De Jesus AMP, Correia JAFO (2008) Fatigue assessment of riveted railway bridge connections. Part II: numerical investigations. In: 7th international conference on steel bridges steel bridges, Guimarães, pp II:339–48
9. Correia JAFO, de Jesus AMP, da Silva ALL, da Silva JFN (2010) A procedure to derive probabilistic fatigue strength data for riveted joints. In: The fifth international conference on bridge maintenance, safety and management (IABMAS2010), Philadelphia, Pennsylvania, USA; 11–15 July 15 2010
10. De Jesus AMP, Pinto H, Fernández-Canteli A, Castillo E, Correia JAFO (2010) Fatigue assessment of a riveted shear splice based on a probabilistic model. Int J Fatigue 32:453–62
11. Jorge RN, Ribeiro A, De Jesus AMP, Figueiredo M, Castro PT, Fernandes AA (2006) Ponte Eiffel – Viana do Castelo, Avaliação de resultados de programa experimental (in Portuguese). Research Report, IDMEC/FEUP
12. Fernandes AA, Castro PT, Figueiredo M, Oliveira F (2004). Structural integrity evaluation of highway riveted bridges. In: Watanabe E et al (eds) Proceedings on international conference on bridge maintenance, safety and management (IABMAS'04), Kyoto, 18–22 October 2004
13. Silva ALL (2009) Fatigue behaviour of a typical material from ancient metallic riveted bridges, under constant and variable amplitude loading. MSc thesis, University of Trás-os-Montes and Alto Douro, Portugal (in Portuguese), 271 p
14. Silva JFN (2009) Comparison of the fatigue behavior between riveted and bolted connections. MSc thesis, University of Trás-os-Montes and Alto Douro, Portugal (in Portuguese), 249 p
15. Figueiredo M, De Jesus AMP, Pereira H, Ribeiro A, Natal R, Moreira P, Castro PT, Fernandes AA (2004) Avaliação da integridade estrutural da Ponte do Pinhão (in Portuguese). Research Report, IDMEC/FEUP
16. Silva T, Fernandes AA (2006) Ponte Trezói – Estudo do comportamento à fadiga de ligações rebitadas em pontes metálicas antigas (in Portuguese). Research Report, FEUP/DEMEGI
17. Correia JAFO. Development of procedures for fatigue life prediction of riveted connections. MSc. Thesis, University of Trás-os-Montes and Alto Douro, Portugal (in Portuguese), 2008. 143 p
18. De Jesus AMP, Silva ALL, Figueiredo MV, Correia JAFO, Ribeiro AS, Fernandes AA (2010) Strain-life and crack propagation fatigue data from several Portuguese old metallic riveted bridges. Eng Fail Anal 17:1495–1499
19. NP EN 10002-1 (2006): Materiais metálicos. Ensaio de tracção. Parte 1: Método de ensaio à temperatura ambiente. Instituto Português da Qualidade (IPQ), Lisboa
20. Chen H, Grondin GI, Driver RG (2005) Fatigue resistance oh high performance steel. Structural Engineering Report No. 258, University of Alberta, Canada
21. ASTM—American Society for Testing and Materials (1998) ASTM E606-92: standard practice for strain controlled fatigue testing. In: Annual book of ASTM standards, part 10, pp 557–71
22. Morrow JD (1965) Cyclic plastic strain energy and fatigue of metals. Int Frict Damp Cyclic Plast ASTM STP 378:45–87
23. Ramberg W, Osgood WR (1943) Description of stress–strain curves by three parameters. NACA tech. note no. 902
24. Coffin LF (1954) A study of the effects of the cyclic thermal stresses on a ductile metal. Trans ASME 76:931–50
25. Manson SS (1954) Behaviour of materials under conditions of thermal stress, NACA TN-2933. National Advisory Committee for Aeronautics
26. Basquin OH (1910) The exponential law of endurance tests. Proc Am Soc Test Mater 10:625–30

27. ASTM—American Society for Testing and Materials (1999) ASTM E647: standard test method for measurement of fatigue crack growth rates. In: Annual book of ASTM standards, vol. 03.01, pp 591–630. ASTM—American Society for Testing and Materials, West Conshohocken, PA
28. Fernandes AA, De Jesus AMP, Silva ALL, Correia JAFO (2012) Retrofitting of old riveted Portuguese bridges. Past and current remnant life assessment research. In: 15th international conference on experimental mechanics (ICEM15); Proceedings on international conference, Porto, Portugal, 22–27 July 2012
29. Paris PC, Erdogan F (1963) A critical analysis of crack propagation laws. Trans ASME Ser E: J Basic Eng 85:528–34
30. APK—Association pour la Promotion de l'Enseignement de la Construction Acier (1996) Construction métallique et mixte acier-béton: Calcul et dimensionnement selon les Eurocodes 3 et 4. Éditions Eyrolles, Paris
31. Branco CM, Fernandes AA, De Catro PMST (1986) Fadiga de Estruturas Soldadas. Fundação Calouste Gulbenkian
32. Weibull W (1961) Fatigue testing and analysis of results. Pergamon Press LTD, London
33. ASTM—American Society for Testing and Materials (2004) ASTM E739-91: standard practice for statistical analysis of linear or linearized stress-life and strain-life fatigue data. In: Annual book of ASTM standards, vol. 03.01, pp 1–7
34. Taras A, Greiner R (2010) Statistical background to the proposed fatigue class catalogue for riveted components. Report: Contribution to WG6.1—assessment of existing steel structures, ECCS TC6—2010, Spring Meeting—Lausanne, March 22–23 2010
35. Glinka G (1985) A notch stress-strain analysis approach to fatigue crack growth. Eng Fract Mech 21:245–261
36. Peeker E, Niemi E (1999) Fatigue crack propagation model based on a local strain approach. J Constr Steel Res 49:139–155
37. Noroozi AH, Glinka G, Lambert S (2005) A two parameter driving force for fatigue crack growth analysis. Int J Fatigue 27:1277–1296
38. Smith KN, Watson P, Topper TH (1970) A stress-strain function for the fatigue of metals. J Mater 5(4):767–78
39. Walker K (1970) The effect of stress ratio during crack propagation and fatigue for 2024-T3 and 7075-T6 aluminum, effects of environment and complex load histories on fatigue life, ASTM STP 462. American Society for Testing and Materials

The manufacturer's authorised representative in the EU is Springer Nature Customer Service Centre GmbH, Europaplatz 3, 69115 Heidelberg, Germany. If you have any concerns regarding our products, please contact ProductSafety@springernature.com

Printed and bound by CPI Group (UK) Ltd, Croydon, CR0 4YY
28/09/2026
02224073-0003